# SUCCESS BY DESIGN

Graphic Design by Daniel Pogacs, Great job! Thanks!

My name is Lewis Daniels.

Let me give you a quick introduction to myself. I hadn't really planned on writing this book. I've previously had two books published by multinational publishing firms, Boonier and Wiley. Those books were about trading; you might therefore ask yourself the question: why would a professional trader now write a book about psychology?

Well, the answer is pretty simple.

Over the 25 years I have been trading, I realised psychology is the number one skill that many traders fail to appreciate in terms of its true power over them. Many traders seem to assume it's all about the strategy or the win rate, not attributing success to the shift in mindset that is required to obtain wealth — or even just to manage stress.

This is not a trading psychology book, not a book about airy-fairy hocus-pocus, or about pushing religion or other beliefs. This is years and years of research, experience and a sudden realisation that what I was actually searching for is, in fact, connected.

My background is a little unconventional, to say the least. When I first left school at 15, I started out as an engineer and used my money from that, whilst still living with my parents, to trade. I had learnt about trading in school.

Another odd story, actually: my history teacher, who wanted to escape the rat race and become a trader, had wangled a school trip from Wales (in the U.K.) to Wall Street in New York. For me, after seeing the trading floor, my pathway was set.

Years later, I find myself living in the French countryside after going through several types of businesses, from beauty salons to making military rapid deployment systems, to technology for banking and payments, and investments in several sectors. I have patents to my name and traditionally published books. However, I still tell people I am a welder.

What this diverse journey has given me is a very unique view of the world.

As a consultant, I help individuals and organisations with trading education, business and life coaching, non-executive and investment requirements.

What that means for this book is an opportunity to speak freely and openly about the components that I now see as interlinked. They were not obvious at first, but they came together — at least for me, anyway — and then it clicked.

I had planned on looking at traditional publishing for this release and decided to go against the traditional route, allowing more freedom to say exactly what I mean.

I hope you find this enlightening and that it motivates you to integrate these powerful insights into your journey.

Zack — This one's for the life you deserve to build.

Angharad – My AMAZING wife, thank you for your patience while I chased questions, and for believing in the life we're building together.

Love you both.

# Success by Design

## Introduction

The idea behind the title of the book is simple. You are capable of designing your own success.

As I said in my brief introduction, some things in life are simply interlinked; you might not realise it yet. I certainly didn't. That realisation took me years, countless books, an embarrassing number of late-night rabbit holes, and probably a million YouTube videos.

Over the years, I've trained thousands of students in the art of trading. One question comes up repeatedly:

"Lewis, why do you bother teaching? If you can trade, why not just trade and disappear?"

The truth is simple: mentoring new traders who are genuinely eager to learn is incredibly rewarding. There is nothing quite like watching that moment when everything clicks—when the fog lifts, the realisation hits, and you see it in their messages or hear it in their voice. The lights turn on.

That feeling only grows when a student comes back to me and says something like:

"Thank you, I just made $10,000 on a trade."

or

“I can’t believe it—I’ve just had my first $100,000 year.”

Those sentences are not about the money for me. They’re about proof. Proof that someone has taken concepts, discipline, and mindset, and actually applied them. Proof that the connection between knowledge and action has finally been made.

In my Discord server—our educational platform where we focus on teaching trading—we have a dedicated channel called ‘why I like it here.’ It’s a simple idea: students write, in their own words, what they appreciate about the community, the teaching, or even just how their thinking has changed.

Reading through that channel gives me a feeling that is very hard to replicate anywhere else. That’s my ‘why.’ That’s a major driving force behind this book.

To be clear about something from the outset: I am not religious. In fact, quite the opposite. I am also not especially interested in conforming to social expectations about how I ‘should’ think, behave, or present myself. As I’m writing this, I live in a remote area of rural France. My nearest neighbour is over a mile away. I can step outside and hear nothing but the wind, the birds, and the occasional tractor.

I’m sharing this to set a context: my perspectives and thought processes do not always follow conventional or academic lines. I have spent a long time studying the various aspects that I'll discuss in each chapter of this book, but what I'm giving you here is not a university lecture or a scientific paper. It is my view and my opinion about how these different worlds collide, intersect, and ultimately connect.

My approach to joining the dots between topics and disciplines is, unsurprisingly, also unconventional.

As a trader, one thing became blindingly obvious early on: psychology is not just 'part' of trading—it is trading. You can have the sharpest strategy in the world, the best indicators, the cleverest algorithm, and a backtest that looks like a dream. It's all meaningless if you cannot manage your emotions in real time.

You need to be able to sit through a winning trade without sabotaging yourself because you're scared to give back profit. You need to be able to withstand drawdowns without panicking and closing positions just before the move you were waiting for finally happens. For many traders, it's not the chart that breaks them—it's their own mind.

That's why I find it amusing when people throw out vague comments like:

"Just fix your psychology, and you'll be fine."

It sounds nice. It's simple. It's also utterly useless.

There's a lot more to it than that. Saying "fix your psychology" is like telling someone building a house, "Just make sure it's stable." Sure, thanks. But how?

This book is not a trading psychology book in the traditional sense. It's not full of airy-fairy, vague, self-help slogans, nor is it an attempt to push religion or any other belief system on you. Instead, it is the result of years of experience across very different fields, and the sudden realisation that the things I'd been searching for—answers, patterns, advantages—were actually connected in ways I hadn't expected.

Let me give you a simple example.

In trading, concepts like Fibonacci retracement techniques are incredibly powerful tools. One of the key levels, 61.8%, is something traders lean on heavily when assessing potential support, resistance, or reversal points. Now, here is where it becomes interesting: that same ratio—approximately 61.8%—appears all over the natural world and in human creations.

Ancient monuments. Famous works of art. The spirals of shells you find on the beach. The structure of certain plants. The proportions of buildings and even faces that we find 'pleasing.' All of them display Fibonacci relationships. The same numbers, the same ratios, repeated again and again.

Yet if you asked a random selection of people on the street about Fibonacci or the golden ratio, more than

60% of them wouldn't have a clue what you were talking about. That gap between what's happening all around us and what we are aware of is one of the themes that fascinates me.

In this book, I will cover Fibonacci, but not just from a "how to use this in trading" perspective. You'll understand its significance beyond finance—how it shows up in nature, design, and even in how we perceive beauty and balance. My interest in these subjects goes far beyond charts and candlesticks.

Trading is only one piece of my story.

Alongside my trading career, I owned a subcontract manufacturing business in the U.K., primarily producing circuit boards for other companies. Some of these were for military projects, others for gaming machines, and others for medical systems. Completely different industries, all fundamentally dependent on electronics and precision.

In 2009, I filed a patent for a unique invention. That journey pulled together engineering, electronics, problem-solving, and an ability to see past what everyone else considered 'obvious.' Later, I moved on to another project—an electric generator that combined mechanical and electronic components. Because I had my own manufacturing facility, I could do more than just sketch ideas. I could physically test them.

Again, different disciplines, different areas of knowledge... but when you see how they fit together, something new becomes possible. This pattern—connecting the 'unrelated'—is at the heart of this book.

This kind of interdisciplinary thinking is essential, yet our world isn't really set up to encourage it.

An electrical engineer is trained to think like an electrical engineer. A car mechanic works on cars and, when a pipe leaks, calls a plumber. A software developer might be brilliant at code but completely lost if you hand them a piece of wood and a saw. None of this is wrong—but it is limiting.

Our education system, from an early age, reinforces silos and categories. We are told what belongs where, and what we are 'allowed' to be good at.

From childhood, many of us are conditioned with one word: "No."

"No, you can't do that."

"No, you can't have that."

"No, it's bedtime."

"No, that's not how it's done."

We are gently but relentlessly nudged into lanes: academic or practical, artistic or technical, this subject or that one. You're a 'math person' or a 'creative.' Rarely both. Rarely encouraged to see how one can enhance the other.

Fast forward to 2022. I purchased an old farmhouse in France. A total blank canvas. Stone walls, beams, fields—the kind of place people imagine when they say they want to 'escape' modern life.

Many of my French friends assumed I was just a "tech guy." After all, I invest in technology, I trade financial markets, and I spend a lot of time in front of a screen. Fair assumption. But what they didn't realise is that before any of that, I worked as a welder. I ran a manufacturing facility. I renovated a beauty salon from the ground up. I oversaw the construction of my own home back in Wales.

When that house was being built, I didn't just sit back and wait for the keys. I was on-site, asking questions and learning. Why use that type of concrete there? Why run the wiring this way instead of that way? Why that tile adhesive? Why that spacing? That angle? I treated the whole build as a live, practical course in construction.

I absorbed everything like a sponge.

Fifteen years later, standing in an old French farmhouse that needed everything done.

I realised how valuable that curiosity had been. Now I had this blank canvas, and I could either outsource everything or I could lean into the weird set of skills I'd accumulated over the years.

One of the projects I decided to tackle myself was building a one-of-a-kind bed out of a sycamore tree. Not buying timber from a shop—actually working with the tree, shaping it, turning it into something functional and beautiful. When I showed the finished piece to a few friends, the reaction was almost identical every time:

"You can't do that—you're a tech guy."

And that, in a sentence, is the problem many of us have. We define ourselves with one label, and then we live inside the walls that label builds around us.

Then they would see something I built and say.

"Why are you not making these for other people? This isn't just furniture; this is art."

My point in telling you all this is that the things you're able to connect in your own life are not always obvious. From the outside, my friends often joke that I'm an alien. They can't get their heads around how someone can solder delicate electronic components to fix an iPhone, build their own bathroom from scratch, and still spend their day tracking global economic events and trading markets.

To me, it's not strange at all. It's just following curiosity wherever it leads.

I'm also a keen scuba diver, and many years ago I took a course on chart work and positioning—learning how to read and use nautical charts, take bearings, calculate positions, and navigate underwater and at sea. At the time, I saw it as something purely related to diving and boating. A single-purpose skill set.

Looking back now, I can't believe how many of those skills and concepts I've applied to other areas of life. The same logic appears when you're running a CNC machine, when you're laying out tiles on a floor, when you're planning a room or aligning a structure. Angles, distances, reference points, margins of error—it's all connected. Different tools, same underlying thinking.

The idea that the skills you learn in one field can—and should—spill over into others. The idea that your brain, your psychology, your habits, and your worldview are not confined to one role or job title, but are part of a bigger, integrated system.

A lot of this will make far more sense as you progress through the chapters. For now, my goal in sharing these stories and examples is simple: I want you to see how crazy some of the connections can be once you stop putting everything in separate boxes.

If you approach this book with an open mind, you may start noticing patterns in your own life: how that seemingly 'random' job you hated actually taught you something you now use; how that hobby you abandoned years ago might still be influencing how you think; how your beliefs about money, work, and success are entangled with things you never questioned.

Stick with me.

For me, when the penny drops—when I suddenly see how two apparently unrelated things are deeply connected—it's one of the most refreshing and energising feelings I know. But I don't think it's enough for me to hand you my conclusions and say, "Here you go."

You need the other pieces.

You need your own experiences.

You need the space to think, to question, and to put your own spin on what I'm going to share.

This book is not here to tell you what to think. It's here to help you see more of the connections that are already there—and, hopefully, to help you use those connections to win. Not just in trading, consulting or coaching or money, but in how you approach your life as a whole.

# Table of Contents

# Chapter 1: The Wealth Mindset

**"A handful of men have become very rich by paying attention to details that most others ignored."**

**Henry Ford**

When I first came across 'The Law of Attraction' – what most people now know as 'The Secret' – I was sceptical. It took several years before my view shifted.

If you've come across it yourself, you'll probably know why. On the surface, it can sound like nonsense: think about money, imagine yourself wealthy, stick a photo of a Lamborghini on your fridge, and somehow the universe will drop it on your driveway. That's how it's usually sold, and that's how I heard it at first. That was my interpretation at the start. It felt like wishful thinking dressed up as wisdom.

Over time, and through a lot of trial, error and observation—both in my own life and in the lives of traders and business owners I've worked with—I realised something important:

It isn't just about believing you will be rich and then it magically happens. It's actually something much deeper than that.

I chose the quote at the start of this chapter to emphasise this point. It's not about positive visualisation while you eat your toast in the morning. It's not about standing in front of a mirror chanting affirmations with no conviction behind them. The people who truly achieve wealth, in any field, have one thing in common:

They are true believers. There is ZERO doubt.

You might say, "Well, that's easy when you have money. Of course, they believe once they've made it." And that's exactly the trap that keeps most people stuck.

I heard something recently that put this into words in a way that really landed for me, even though on paper it was obvious:

People often say, "I'll believe it when it happens."

On the surface, that sounds sensible. Evidence first, belief second. But buried in that statement is the problem: it implies doubt. It says, "I don't truly believe now; I'm waiting for proof." If you carry that attitude into money, business, trading—anything, really—you've already set a limit on what you're willing to commit to and what you are willing to believe is possible for you.

This is the difference.

As you go through the rest of this book, I will keep pointing back to this. You'll start to see the true difference between lazily thinking about something and half-hoping it appears like an Amazon parcel at your door... and actually believing it with every ounce of your being. Living from that belief. Making decisions from that belief. Acting as if the outcome is done and you're simply moving towards it.

That is what I mean by a wealth mindset.

The concept itself is not new. The wealth mindset has been covered in hundreds, if not thousands, of books. One of the most famous is Napoleon Hill's Think and Grow Rich. I've read it. I've read many others like it. There is a huge amount of value in those books, and if you're anything like me, you could easily spend a decade reading and studying the psychology of success, money, mindset—and still miss the critical part.

That tiny switch.

It seems to elude many people forever. They consume content, highlight passages, nod their heads, feel motivated, and yet nothing fundamental changes. Their life looks the same five years later—maybe with more quotes saved on their phone and a slightly bigger book collection.

Why?

Because it is not as simple as reading one book or watching one video and suddenly having 'the secret' to wealth. You can read a thousand of them, be a lot wiser, and still not flip that switch in your head that changes everything.

Look at sports personalities. Boxers are maybe the best example. In the weeks leading up to a big fight, they trash-talk each other relentlessly. Both of them are 'the best in the world'. Both of them 'can't be beaten'. Both of them 'won't be beaten'. If you took those words literally, you'd have to laugh—obviously, both can't be right.

Now, do you think that's just a clever marketing trick to sell more tickets and Pay-Per-View subscriptions? Of course, there's a commercial element. But there's also something far more important going on beneath the surface.

What they're actually doing is convincing themselves. They're programming their own mind. They're teaching their brain to deliver its best possible performance on the night. The trash talk is as much internal as it is external. They are removing doubt. Because doubt in the ring is dangerous. Hesitation in a split-second situation can get you knocked out.

I'll cover this performance element in more depth in a later chapter, especially how it ties into business, money and decision-making under pressure. For now, the point is this:

The mindset is the trigger.

The mindset is the key.

And if you can flip that internal switch in your brain to allow the wealth mindset to operate efficiently, you will be shocked at what you can achieve.

A lot of people will tell you that a 'wealth mindset' means living below your means, cutting back on coffee, driving an old car, and slowly building up a nest egg. And yes, discipline with money is important. I'm not telling you to max out credit cards on watches and cars 'for the vibe'.

But let's be clear: this popular idea of extreme frugality as a wealth mindset is, in many cases, just a security net. It often reveals doubt in your own capabilities rather than confidence. It says, "I don't trust myself to create more, so I'll shrink everything down and cling to what I have."

You see where this is going?

What about persistence and resilience? Open almost any self-help or success book, and you'll read about how 'you must be resilient', how 'persistence is vital', how you 'should never give up'.

Not only is it boring, but on its own, it's also inaccurate and incomplete.

You can be the most persistent person in the world, running full speed in the wrong direction. You can be incredibly resilient in holding on to beliefs that are actively sabotaging you. Persistence without clarity of belief is just stubbornness.

The deeper truth behind a wealthy mindset is this:

It's all about removing doubt.

You need to strip doubt away from the outcome you're moving toward. Not from every detail or every step of the journey—you will still have uncertainty there—but from the core belief that you will, ultimately, get where you're going.

Jeff Bezos didn't sit down to build Amazon and say, "Ah, there are libraries and bookshops already; this might fail." He wasn't blind to risk. He knew the odds. But at a fundamental level, his conviction was that this idea could work and would be built. He operated from that belief.

Mike Tyson did not walk into the ring thinking, "I might lose this one." He trained with the mindset that he was the most dangerous man on the planet. Was it arrogant? Maybe. Was it effective? Definitely.

As a rugby fan, I've watched the New Zealand All Blacks for years. Many times, you could feel that they'd already won the match mentally before they even stepped onto the field. The opposition often turned up half-beaten. That's the power of collective belief—of a mindset installed and reinforced over years.

Now you might be thinking, "That's all very well for billionaires and world-class athletes, but what about normal people? What about me? It's not as easy as just believing, especially when bills need paying, you've got responsibilities, and setbacks are part of everyday life."

You're right: it's not 'easy'. But that doesn't make it complicated.

Setbacks are part of the process, not evidence that the belief was wrong. When New Zealand lose a match, they don't burn the jersey and decide rugby isn't for them. They review, adjust, and move on. They do not let a loss rewrite their identity.

Setbacks in your financial life, or in business—these are your losing matches. Feel them, learn from them, then get over them and move on. If every setback kills your belief, you never had belief to begin with. You had a preference. A wish. A 'nice to have'.

Most books that claim to help you 'shift your mindset' are obsessed with the outcomes: more money, better habits, nicer things. But the mindset itself—what actually happens internally when someone genuinely flips that switch—is rarely covered in a meaningful way.

This brings us back to the Law of Attraction.

When you say to yourself, "I want to be rich," all you've done is place the outcome in the future. It's somewhere ahead of you, out of reach. Your subconscious mind hears that and says, "Right, so we're not rich yet. We're just wanting." So it keeps you in that loop. Wanting. Chasing. Dreaming.

You become an excellent wisher, a professional visualiser – but nothing significant changes.

The real mindset shift is much more radical than that. It requires you to live in the end result now. Not to fake it with empty affirmations, but to begin thinking, behaving and making decisions as if the outcome is a done deal and you're simply catching up in real time.

This is where most people struggle. And understandably so.

I know it's hard to think like a millionaire when your bank balance looks sickly. I've been there. I know what it's like to run out of money before the end of the month, and to feel like you're a million miles away from where you want to be. But that's precisely why the shift is so powerful. It isn't about pretending your circumstances are different. It's about changing the way you relate to them.

The whole idea is to make conscious changes that bring a better overall emotional state. Not fake happiness, but a more empowered position.

Take a very simple example. You wake up and say, "Today is going to be a good day," then immediately pull the pillow back over your head for an extra five minutes because you can't be bothered getting up.

That is contradictory.

Your subconscious doesn't respond to what you say as much as it responds to what you consistently do and feel.

It sends mixed messages to your brain. You're saying one thing and acting in a completely different way. The small actions, thoughts, feelings and the tiny daily choices—those are the signals it listens to.

The small changes lead to bigger outcomes. It isn't luck. It's a shift in mindset, repeated often enough that your brain has no choice but to reorganise around it.

The challenge is that when you hear phrases like "The Secret" or "Law of Attraction," your brain often files them under 'magic', 'nonsense' or 'too good to be true.' That automatic dismissal flips a switch in your subconscious mind. In that instant, you're anchored back to your old beliefs:

"Money is hard to obtain."

"People like me don't get rich."

"It's all rigged anyway."

So you go back to the day job, mentally and emotionally. Even if you're physically trying something new, internally you're still running the old program.

Let's imagine you get past that first barrier. You decide, "No, I'm going to believe this. I'm going to take it seriously." What happens next for most people is one of two things.

First possibility: you run straight into a setback, as you inevitably will at some point.

You lose money on a trade, your business idea flops, your 'big break' falls through. Instantly, the old narrative springs back up: "See? I knew it. This stuff doesn't work. It's all rubbish." The belief dies on contact with reality.

Second possibility: you paint the mental picture so far into the future and so far from your current reality that your brain quietly refuses to engage. You imagine yourself as a billionaire in five years with zero intermediate steps. It's so far removed from your current experience that your mind basically says, "Nice fantasy. Not for us."

In both cases, the result is the same: no genuine, sustainable shift.

This is only the first chapter, so some of what I'm saying now might feel incomplete or even slightly frustrating. That's deliberate. I don't want to pretend you can read a handful of paragraphs and rewire a lifetime of conditioning. You will need the rest of the book to see how all the pieces fit together.

For now, here's what I want you to take away from this chapter on the wealth mindset:

It is not wishful thinking.

It is not blind optimism.

It is not frugality dressed up as wisdom.

It is not about pretending setbacks don't hurt.

The wealth mindset is about removing doubt at the core level of who you believe you are and what you believe is possible for you. It's about moving from "I hope this works" to "This is happening—I'm simply aligning myself with it."

As you move through the following chapters, you'll see how psychology, patterns like Fibonacci, trading behaviour, and even experiences outside of finance all tie back into this central idea. You'll start to notice where you're quietly sabotaging yourself with doubt, even in the way you phrase your goals and talk about your own future. For now, just sit with this question:
If you removed all doubt in the end result—how differently would you think, speak, and act today?

Answer that honestly, and you've just taken your first real step into a genuine wealth mindset.

# Chapter 2: The Convergence of Faith and Fortune

**"What you think, you create. What you feel, you attract. What you imagine, you become."**

**Deepak Chopra**

I am the least religious person I know.

Let's get that out of the way up front, because if you're anything like me, your guard probably goes up the moment someone starts talking about faith, God, or religion in the context of money.

But my research has led me down this path in several different ways, and I'd be lying if I said I could just ignore it. So let's throw it straight out there:

What if the Bible is not purely a religious story, but a set of instructions?

There's a line in Exodus where God says to Moses:

"I AM WHO I AM."

And he said, "Say this to the people of Israel: 'I AM has sent me to you.'"

I'm not here to give you a Bible study. That's not my world. But when I came across that, something about it bothered me—in a good way. It stuck in my head. "I AM" is a powerful statement. It's definitive. It's present tense.

It isn't "I hope to be", "I might be one day", or "I used to be". It's identity. Final.

And then a thought hit me:

What if this isn't just about a God talking to Moses?

What if it's also a coded lesson about how we use "I am"?

"I am capable."

"I am useless."

"I am broke."

"I am wealthy."

"I am unlucky."

"I am disciplined."

Every time you make one of those statements, you're issuing an instruction to your subconscious mind. You're declaring something about who you are, not just what you want. That's a big difference. This will make a lot more sense in later chapters.

If you do some research for yourself (and I encourage you to, because I'm not going deep down that rabbit hole here), you'll find something interesting: a very high percentage of the world's centimillionaires and billionaires are religious or claim some form of spiritual belief.

That doesn't mean every religious person is wealthy, far from it. It also doesn't mean every wealthy person is religious in the traditional sense. But the overlap is noticeable enough to be worth paying attention to.

That got me thinking.

What if religious belief, at least in part, is simply a structured way to apply a standard of living and a standard of thinking? What if, underneath all the rituals and stories, it's a system for programming belief and identity?

"I AM" is a powerful statement. What if this really means "YOU ARE"... or "WE ARE"?

Take Henry Ford's famous quote:

"Whether you think you can, or you think you can't – you're right."

Strip away the religious language, and you see the same principle:

What you accept as true about yourself defines your limits.

What you believe you are capable of determines how you behave.

How you behave repeatedly becomes your reality.

So what if, instead of always looking outside for some external saviour, we start to look inside? Back to Chapter One—it's the mindset shift again.

Now let's push this further.

What if religion, in its purest and least corrupted form, is trying to tell us to look inside ourselves? What if the stories aren't just history or mythology, but metaphors for what's possible when belief and action align?

We have the ability to slay giants—David and Goliath style. Not literal three-metre warriors, but your debts, your fears, your lack of education, your circumstances.

We have the ability to part seas—maybe not physically, but to do things that look impossible from where we're standing right now.

These don't have to be 'miracles' in the supernatural sense. They could just as easily be descriptions of what a human being can do when they have absolute conviction, absolute faith—especially in themselves.

During my research, a lot of roads started pointing back to the same places. Even when I tried taking the long way around—starting from science, psychology, trading, philosophy—they eventually circled back.

## Symbols. Structures. Ratios. Fibonacci.

As science advances, we're finding structures and images at an atomic and subatomic level that have similar, or even direct, comparison to religious symbols and sacred geometry. Spirals, ratios, repeating patterns. The golden ratio, 1.618, pops up in nature, architecture, and art—and how 61.8% shows up in Fibonacci retracements in trading.

There's an experiment online where a drop of water is placed on a flat plate, and sound or music is played. As each note is hit, the water shifts its shape. Specific frequencies create specific patterns.

Similarly, there's another experiment using sand or rice on a vibrating plate. Change the frequency, and the grains rearrange into beautifully ordered shapes: geometric mandalas that look suspiciously like motifs you see in temples and religious art.

Sound creating structure. Vibration creating form.

Go and research the Mandelbrot set when you have time. A simple mathematical formula, repeated over and over, produces an infinitely complex, self-similar pattern. Zoom in, and you see the same shape repeating at different scales. It's a visual demonstration of how simple rules can generate endless complexity.

Now stack that against religious imagery, symbols, and the idea that "in the beginning was the Word" (sound, vibration, expression). Start connecting that to the idea that your words—especially your "I AM" statements—carry weight.

Is it really that far-fetched to think there might be a link between what we say, what we believe, and the structure our life takes on?

Here's where it gets even more interesting.

It is said that the Vatican has secret libraries, filled with texts not accessible to the public. Then there's the famous fire at the Library of Alexandria, which supposedly destroyed an enormous amount of ancient knowledge. Whether you believe every detail of those stories or not, ask yourself this:

Do you think there might be a reason for hidden or lost knowledge?

What if there were an instruction manual for life, scattered across religion, philosophy, geometry, biology, and psychology?

What if religion, or secret societies such as the Illuminati or the Freemasons, have access—or at least claim to have access—to knowledge or tools that could literally change the way we see the world?

I'm not asking you to put on a tin-foil hat here. Let's put the conspiracy theories aside for a moment. Forget hidden bunkers and clandestine meetings. Even at the most surface level, without going underground or mystical:

What if religion is a roadmap?

What if it's trying—sometimes clumsily, sometimes clearly—to assist you in shifting your mindset?

What if instead of seeing these stories purely as historical or supernatural events, we see them as a mirror?

By that, I mean: what we put in, we get out.

Back to the wealth mindset from Chapter One. If you believe you can beat a Goliath—your debt, your fear, your lack of education, your circumstances—you are far more likely to do it. If you believe you can't, you're almost guaranteeing your own defeat.

When you look in the mirror and smile, the mirror smiles back. When you scowl, you don't see a happy face staring back at you—you see your frown, reflected, multiplied.

Your life often works the same way.

Now, I might be way off track here. I'm happy to admit that. But let's assume, for a moment, that I'm onto something.

The belief in a higher power—call it God, the universe, Source, whatever you like—may actually be a structured way of accessing faith. And that faith, in practice, is something you assimilate into yourself.

In other words: the power is not out there, somewhere in the clouds. It's being installed in here, in you.

So while some people see verses like "put on the full armour of God" as an expression of faith and protection from evil, let's remove the word 'God' for a moment and see what happens.

"Put on the full armour of..."

Discipline, Belief, Focus, Self-respect, Vision.

What if, instead of clothing yourself in religion, you are clothing yourself in a state of mind?

You might have heard the saying: "Sticks and stones may break my bones, but words can never hurt me."

It sounds tough. But it's a lie.

Words can and do hurt. Not just the ones directed at you by others—but especially the ones you direct at yourself. The constant internal drip feed:

"I'm not good enough."

"I've always been bad with money."

"I can't do what they do."

"It's too late for me."

These are like microscopic cuts you inflict on your own belief system, day after day, year after year. They weaken your armour from the inside.

Now bring the two together.

Put on your protective armour—and be conscious of the words you accept and the identity you wear.

This is where belief in yourself and religion start to merge.

Whether you kneel in a church, sit in meditation, or simply close your eyes before a big decision and say, "I've got this," you're doing a similar thing:

You are choosing a state of faith over a state of fear.

Think about athletes before a major event. Many of them pray. Some cross themselves. Some point to the sky. Some repeat mantras. Others have seemingly weird rituals—lucky socks, specific routines.

Are they all religious? No. But they're all doing a version of the same thing: centring themselves, activating belief, and connecting to something bigger than their fear.

When you strip away the surface-level differences, prayer, meditation, visualisation, and strong self-talk are cousins. There are ways of focusing your attention and directing your belief.

Now let's talk about fortune. Money. Wealth. Because this book isn't just about abstract spiritual ideas—it's about how these ideas tie into business, and building a life that actually works.

If you look closely at many wealthy individuals—not the lottery winners, but the builders, the repeat performers—you'll notice patterns.

They often have some form of structured belief system. It may be religious, spiritual, philosophical, or even just a personal code. They operate from a strong internal sense of "I am": "I am a creator", "I am a winner", "I am relentless", "I am valuable". And they generally act as if the world is, on balance, something they can influence. They don't see themselves purely as victims of fate.

That last point is crucial. Psychologists call it 'Locus of Control'—whether you believe life happens to you or through you and with you.

If your entire mindset is external—"God will fix it," "the government will fix it," "the market will fix it," "a miracle will save me"—you are outsourcing your power. You're effectively saying, "I am not responsible. Something else is."

On the flip side, when you recognise that faith, properly understood, is not passive but active, things shift.

"I have faith" becomes:

"I trust that if I move, things will move with me."

"I trust that if I commit, doors will appear."

"I trust that if I walk, the path will form under my feet."

And crucially: "I trust myself to follow through."

This is where faith and fortune begin to converge.

Money, wealth, and success in any meaningful form require you to tolerate uncertainty. They require you to act without guarantees. To invest. To risk. To push chips into the middle of the table when other people are folding.

If you wait for certainty, you'll wait your entire life.

Religious language talks about "walking by faith, not by sight." Translate that into trader-speak or entrepreneurial language, and you're basically saying:

"You make decisions with incomplete information. You act without every variable locked in. You place trades without knowing the outcome in advance."

You don't know the exact path. But you move anyway.

Building wealth is exactly like that. If you need everything to be 100% clear before you act, you'll never start. And if you do start, you'll bail at the first sign of trouble.

This is why so many people read about wealth, watch videos on mindset, attend seminars... and then freeze. They've filled their head with ideas but never cultivated actual faith—especially faith in themselves.

They've outsourced faith to a theory, a guru, a book, or a magic law. And when reality doesn't immediately match the promise, they collapse internally.

Let me be very direct here:

Faith—even religious faith—without personal responsibility is useless for building fortune. Faith with personal responsibility is explosive.

Think of it as a two-part system.

First, belief in a higher order, pattern, or structure to life. Call it God, the universe, natural law, whatever you like. This gives you a sense that your efforts matter, that there is some kind of coherence to things.

Second, belief in yourself as an active participant. This is the "I AM" piece. Not "I am waiting," but "I am a creator, I am an effective agent, I am capable of change and progress."

Combine those two, and you get a mindset that says: "I am supported—but I am also responsible."

If you're like me and you're not particularly religious, you don't have to adopt anyone's doctrine to benefit from this. You don't need to suddenly start quoting scripture to operate with faith.

You just need to grasp and apply the underlying principle:

What you repeatedly say to yourself becomes what you believe.

What you believe determines what you attempt.

What you attempt (and persist with) largely determines what you experience.

In trading terms, if you internalise "I am a losing trader," you will find ways to prove that true. You'll hesitate, overtrade, revenge trade, and sabotage good setups. You will unconsciously return to the identity you've accepted.

If you adopt "I am a professional trader learning to master my craft," your behaviour changes. You start acting like that person: you protect capital, follow rules, review your trades, and control risk. The results follow the identity, not the other way around.

Now bring that back to "I AM WHO I AM."

What if that isn't just a divine statement? What if it's a template?

"I AM..."

—and you fill in the rest every day with your thoughts, words, and actions.

If there were an instruction manual for life, it might start with that. Because once your identity is set, everything else falls into line. Faith just becomes the fuel that keeps that identity operating consistently when reality throws tests your way.

The convergence of faith and fortune, then, looks something like this:

You recognise that belief, in some form, is unavoidable.

You choose to believe you can influence your financial reality.

You consciously shape your "I am" statements instead of repeating old, inherited ones.

You act in alignment with those statements—again and again—especially when it's uncomfortable.

You see setbacks not as divine punishment or proof that "it's not meant to be", but as tests, feedback, and adjustments.

You don't sit back and wait for a miracle. You become the vehicle through which the 'miracle' happens.

That's what I see again and again in people who build serious wealth. Some wrap it in religious language, some don't. But the mechanism is remarkably similar.

So, where does that leave you?

You don't need to buy into any specific religion to benefit from the core idea of faith. You don't need to attend a place of worship or subscribe to a doctrine to start reshaping your internal statements.

"I am stuck" can become "I am learning."

"I am broke" can become "I am building."

"I am hopeless with money" can become "I am becoming financially competent."

These might sound small or even cheesy, but remember: your subconscious doesn't respond to dramatic speeches—it responds to repetition and alignment. Words backed by action.

When you say, "I am becoming financially competent," and then you sit down, learn, study, track your spending, run simulations, take controlled risks—now your words have teeth. The mirror starts to reflect something different back to you.

In that moment, the line between religious faith and self-faith blurs. You are, in effect, doing what billions of people have done for centuries when they pray, chant, or repeat sacred phrases:

You are directing your consciousness.

You are choosing your focus.

You are betting on a certain version of yourself.

The difference is that you're doing it deliberately, with your eyes open, and you're tying it directly to your financial life.

To wrap this chapter up, I'm not asking you to become religious. I'm not asking you to accept anyone's version of God. I am asking you to recognise the raw power of belief, language, and identity in shaping your financial outcomes.

Faith and fortune are not as far apart as they seem.

The belief that 'something out there' has your back can give you courage.

The belief that 'something in here' is capable can give you action.

Where those two overlap—where you stop outsourcing everything to fate and stop pretending you're powerless—that's where a different level of wealth can begin.

Because if faith is nothing more, at its core, than a deep conviction that what you move toward will, eventually, meet you halfway...

...then the convergence of faith and fortune is simply this:

You move. You believe. You persist.  And you let your "I AM" become the foundation on which your wealth is built.

# Chapter 3: Visualisation and Success

**"In my mind, I've always been an A-list Hollywood superstar. Y'all just didn't know yet."**

**Will Smith**

Isn’t this the same as the wealth mindset?

Well... yes and no.

The difference is in the technique.

Shifting your wealth mindset and building belief in yourself is one part of the equation. The second half is the way you use your imagination—the power of visualisation.

Picture a mood board. In the middle is that super-yacht, your dream. Around it, maybe a house, a car, a few quotes you found on social media. You look at it, feel a little buzz, and then what does your mind do next?

For most people, it goes straight back to the day job.

“Nice dream. Back to reality.”

What about the faith talked about in chapter 2? Well, if dismissal is your automatic response, clearly you don’t really believe in yourself yet. You’ve still got too many giants to face and a sea to cross.

During my own research, I came across the teachings of a man called Neville Goddard (1905–1972). He was a mystic, author, and lecturer who spent his life talking about the power of imagination and consciousness. He hammered the point that belief and visualisation are not just nice ideas; they are the mechanisms through which your assumptions shape your reality.

His philosophy blended new thought and Christian mysticism, but you don't have to buy the labels to see the logic. He wrote books like 'The Power of Awareness', 'Feeling Is the Secret', and 'Awakened Imagination', exploring concepts like the law of assumption and creative visualisation. A lot of what you see in modern self-help and 'manifestation' circles traces straight back to him.

Neville talked about "living in the end."

Instead of visualising what you want and keeping it parked in the future—as a 'want', a someday—you shift into seeing it as already done.

So yes, you can change your wealth mindset, and that will absolutely drive better decision-making and attract more positivity. But the art of living in the end is the next level.

Go back to that picture of the yacht in the middle of your mood board. It isn't enough to just want it. You have to imagine the feeling of owning it. The sea air on your face. The sensation of waves hitting the hull. The sunshine warming your skin. The sound of water and engines, the laughter of friends, the quiet moment alone at sunrise on deck.

You have to see it in your mind as if you are already there.

Here's the key: your mind doesn't really know the difference between imagination and reality.

Research using functional MRI (fMRI) has shown that imagining an action activates many of the same areas of the brain as actually performing that action. Studies on imagined physical movements—for example, athletes visualising their performance—show that similar neural pathways light up as when they execute those movements for real.

Cognitive psychology has also demonstrated that imagined events can become almost indistinguishable from real memories. Elizabeth Loftus's research on false memories shows that people can confidently recall events that never happened. Their brain has accepted the imagined event as 'real enough.'

Thoughts become tangible.

When these things started to make sense to me, it really changed how I saw the whole world.

You might have experienced a much simpler version of this. You think of someone you haven't spoken to in ages… and they call or message you out of the blue. Coincidence? Maybe. But the same principle is at play.

So when you visualise yourself doing the same job for the next 40 years, feeling trapped and resentful, guess what kind of opportunities—if any—you're likely to notice and accept?

Now that you've read this, pay attention the next time you watch interviews with famous actors, athletes, or successful entrepreneurs. Listen carefully. You'll start hearing them say things that line up with this: "I saw it before it happened," "I knew it was going to be mine," "I visualised the win," "I'd run the scene in my head years before it was real."

They're not being poetic. They're describing a practice.

You can see, I hope, where this is going.

When you have true belief in yourself, and then you create powerful visualisations of the future you want, things start to shift. People say "the universe will conform." But most people hear that as wishful thinking. Without realising it, they drift back into old habits and old ways of thinking.

They believe abundance is not for them. Wealth is something in the distant future—if it comes at all. You hear the same phrases over and over: "It's easy for them; they have money," "People like us don't get those chances."

Yet when you listen carefully to the language used by the successful, you'll notice something: they don't just believe in themselves; they believe in their 'future'. They speak as if it's already written.

And here's the crucial bit: did they only start visualising after they became successful?

Absolutely not.

This is why this narrative matters. The road to success is built with authenticity, clarity, and a deep-rooted belief in your potential. When you envision your success as inevitable, your actions begin to align. And when your actions align, opportunities appear that you didn't even know existed.

Let's just take a minute here, as this is a game-changer.

Opportunities appear that you didn't even know existed.

Often, when visualising, you unconsciously can set limitations, trying to set realistic goals, what you think of as practical aspirations, achievable dreams, finding some logical path to how they could become reality, when in truth there are no limitations to your visualisations, your dreams or goals. There are infinite possibilities and opportunities even beyond our comprehension. When you fully submerge in the belief of your visualisation, the universe responds with extraordinary ways to align that desire.

So let's bring this down from theory and talk about how you actually do it.

Start by finding a quiet space where you won't be disturbed. Sit or lie comfortably. Close your eyes. Take a few slow breaths and centre yourself.

Visualisation is not just about seeing; it's about 'feeling'.

Picture your future in vivid detail: the sights, the sounds, the smells, the emotions that go with your success. Don't watch it like a film on a screen—step into it. Be in the scene.

If your visualisation involves that luxurious yacht, go all in. Feel the texture of the polished wood under your bare feet. Hear the gentle slap of waves against the hull and the cry of seabirds. Smell the salt, the sunscreen, the coffee. See the sunlight hitting the water. Most importantly, feel what's happening in your chest—pride, joy, freedom, calm.

Engage as many senses as you can. That's the trick. You're not just visualising the 'want'; you're feeling the emotions as if it's already done.

Then consider setting up a daily visualisation practice.

Consistency is key. Ten to fifteen minutes a day is enough if you're fully engaged. Do it in the morning to set the tone, or in the evening to reset your mind before sleep. Use images, words, or music that resonate with your vision to deepen the experience.

Mood boards can help here, not as 'wish lists', but as anchors—physical representations of your goals that remind you not of what you lack, but of what you're moving into.

Alongside visualisation, bring in affirmations that match your vision. If your goal is to become a successful entrepreneur, you might repeat to yourself, "I am a leader who inspires others," or "My business thrives because I attract the right opportunities," or "I make clear, confident decisions that move me forward."

These lines are not magic spells. They're instructions. They reinforce your belief system and help align your subconscious with your conscious goals.

Practice Gratitude – be grateful for what you already have, what you have already achieved. It's so easy to become complacent. To not appreciate the present and always long for a better tomorrow. So when you're cleaning your teeth every morning, think of something you're grateful for. You might want more out of life, but right now, today, what are you grateful for? Your family? Your health? Your house? Hot water? Not having an alarm? Think of the things both big and small. It might seem irrelevant, but this simple act can fill you with joy, with pride, with happiness, and it can help transform your day into one of productivity and determination. Appreciating what we already have can help us achieve what we really want.

Another powerful tool is journaling.

Write about your desired outcomes as if they have already happened. This supports Neville's idea of living in the end. Describe a day in your life as a successful trader, business owner, or expert in your field. What time do you wake up? What does your environment look like? Who do you talk to? What kinds of problems are you solving? How do you feel as you move through that day?

Journaling does two important things. It forces you to clarify what you actually want, and it reinforces your vision by giving it structure and language.

You might be thinking, "Alright, but how does this tie into real-life actions?"

Visualisation and belief must end in action. Otherwise, it is just daydreaming.

To do this properly, once you've pictured your success clearly, ask yourself one very simple question:

"What one step can I take today that moves me towards this vision?"

Your subconscious will start to nudge you—toward a conversation, a book, a course, a mentor, a change in routine. You might feel a quiet push to network more, refine a skill, save differently, or take a calculated risk.

Trust that guidance, especially when it's pointing you towards things that feel slightly uncomfortable but clearly aligned.

It's crucial to understand: visualisation is not a substitute for hard work.

Mental imagery is powerful, but it has to be tied to effort, persistence, and a willingness to adapt. You will need to get comfortable taking risks and stepping outside your comfort zone. Success is rarely a straight, clean line; it's a winding path full of detours and lessons.

Another key point: be vigilant against negativity and self-doubt.

This is where most people fall off. They visualise for a week or two, feel good for a while, and then something goes wrong. A bill lands. A trade fails. A plan falls apart. Instantly, they're dragged back to the old life, the old thinking. Negativity creeps in, doubt takes root, and just like that, they're back living in the old world.

Visualisation doesn't fail them. They abandon it.

Surround yourself with positive influences—people who uplift you, challenge your limits, and don't mock your ambitions. That support system reinforces your belief in what you can achieve.

Invite accountability too. That doesn't mean announcing your goals to the whole world. But be accountable to yourself. Own your vision. Repeat it. Track it. That alone increases the positive charge behind it.

And remember: be patient with yourself.

Success does not usually come overnight. It's the product of sustained effort, learning from failures, and continually refining your approach. Your visualisations will evolve as you grow. What you picture today may be very different from what you picture in five years—and that's a good sign. It means you're expanding.

As we move further into this book and go deeper into the science and logic behind these principles, you'll see they're not just fluffy ideas. They're rooted in psychological and neurological research that validates their impact. Understanding that foundation will give you even more confidence in using visualisation as a tool, rather than treating it as some mystical trick.

A piece of music, a scent, a memory, all can trigger us to relive emotional states, both good and bad. The intensity that these feelings have on us and our mood, our actions, is no different to the power visualisation can have. We can harness this power to invoke positive change.

By now, I hope you're starting to see the immense potential within yourself through the lens of visualisation.

If you embrace its power fully, you may discover that the barriers you once saw as solid walls are actually just stepping stones—or illusions—on the path to the life you envision.

The universe responds.

Practise living in the end.

Hold onto your dreams. Align your actions with them.

Success isn't just about reaching a destination; it's about walking the path in a way that transforms you.

Visualisation is how you start walking that path in your mind—long before your feet touch the ground.

Elite performers across disciplines share one quiet habit: they see it before they do it. Formula 1 drivers, world-class fighters, record-breaking swimmers like Michael Phelps, even free solo climbers. All use vivid mental rehearsal. The pattern is unmistakable: in the top 1%, visualisation isn't optional. It's part of the work. If you want to live in that tier, it's not that you should visualise. You must.

# Chapter 4: The Science of Attraction

**"You are a vibrational selector. You attract what you are, not what you want. So if you are always thinking about what you lack, you will attract lack."**

Esther Hicks

As we journey deeper into this fascinating thing we call 'success', it's worth lifting the bonnet and looking at the mechanics underneath.

So far, we've talked about wealth mindset, the convergence of faith and fortune, and the power of visualisation. On the surface, those ideas can sound like 'woo-woo', luck, or mysticism. But underneath the language, there are very real mechanisms at work: psychology, neuroscience, biology, quantum physics, body language, and something most people underestimate—momentum.

Reality is not just mystical; it is mechanical. And you are influencing it constantly, whether you realise it or not.

Read the quote at the top of the chapter again.

Let's start simple. No quantum entanglement just yet.

Stand in front of a mirror and force yourself to smile.

If you're in a bad mood, you will feel ridiculous for a second. But hold it. Notice what happens to your face, your eyes, your breathing. Notice how difficult it is to stay in a truly dark, miserable state while your facial muscles are arranged into a genuine smile—not the fake, strained kind, but a real one.

That single action—moving muscles in your face—changes three things at once.

It changes the feedback sent to your brain. It changes the chemicals your brain releases. And if someone else is in the room, it changes how they respond to you.

There's a principle in psychology called the facial feedback hypothesis. It says your facial expression doesn't just reflect your emotion; it can influence it. When you smile, you're not just broadcasting happiness—you're quietly instructing your brain to move in that direction.

Now flip the scene.

Walk into a room with your shoulders slumped, eyes down, brow furrowed. Before you say a word, you're broadcasting something completely different. People's brains are making micro-calculations:

"This person looks defeated."

"Low energy."

"Not confident."

Their behaviour towards you will shift. They may trust you less, engage less, invest less—financially or emotionally.

What changed? Same person. Same history. Same skills. Just expression and posture.

This is the most basic level of the 'science of attraction': you are sending signals all the time—through your face, your tone, your timing, your body language. Those signals alter both your internal chemistry and the way others respond.

It's not magic. It's mechanics.

At the core of this sits something called Neuro-Linguistic Programming, or NLP. Forget the hype and seminars for a moment; at its core, NLP is just a framework for understanding how language, imagery, and feeling combine to shape your reality.

It starts with a blunt idea:

The words you use, the pictures you run in your head, and the feelings you attach to those pictures all combine to create your experienced reality.

If you constantly tell yourself, "I am terrible with money," your brain doesn't treat that as a joke. It treats it as code. A command. It starts filtering the world through that lens.

You don't see opportunities to grow wealth because you 'know' you're terrible with money. You don't seek financial education because 'it won't stick anyway.' You don't negotiate, you don't ask, you don't push, because deep down you've already decided you'll lose.

NLP suggests that if you change the language and the internal pictures, you change the experiences you create and notice. That fits perfectly with the wealth mindset from Chapter One.

When you cultivate an internal dialogue filled with abundance instead of scarcity, you shift your perception.

You start noticing different things, leaning towards different choices, saying yes to opportunities you used to ignore—or couldn't even see.

This is where confidence stops being a fluffy word and becomes something you can observe.

Confidence is not abstract. You can see it. You can hear it.

It's in eye contact.

In tone and pace of voice.

In how someone occupies space.

In how they answer questions.

In how quickly and cleanly they make decisions.

Research shows that confidence strongly influences how others rate your competence—sometimes more than your actual competence in the moment. That doesn't mean you can fake your way through life forever, but it does mean that how you present your belief in yourself is crucial.

When you genuinely believe in yourself, and you've spent time visualising your outcomes until they feel familiar, your body language shifts naturally.

You sit differently.

You hold your head differently.

You walk into a room differently.

You pitch differently.

You don't shrink from questions.

You're not acting. You're aligned. People can feel that.

Take a simple scenario: a meeting with a venture capitalist for your start-up.

Most first-time founders walk into that room almost begging for assistance. Their internal script is: "Please approve of me. Please give me a chance. Please save me." That script leaks out—in their words, their nerves, their posture. The investor smells desperation. They hear uncertainty in every answer. The pitch feels shaky.

That often ends with, "We'll pass," or, "Come back when you have more traction."

Now contrast that with a founder who has been working on the inner game as well as the deck.

They have a strong wealth mindset: "I am building something valuable."

They have visualised this meeting dozens of times.

They feel their future as if it's already in motion.

They don't walk in apologising for asking for money. They walk in offering an opportunity. Their tone, their posture, their presence all say:

"I know where this is going. You can join or not, but this is happening."

Same product? Maybe. Same slides? Possibly. Completely different energy.

That's the first layer of attraction: psychological and behavioural. Visible. Tangible. You can see it in the way people live.

But the science doesn't stop there.

Now, let's zoom right out—and down—to the quantum level.

Quantum mechanics is the branch of physics that deals with the tiniest building blocks of our universe: photons, electrons, and subatomic particles. This is where things stop behaving like billiard balls and start acting like something out of a philosophy seminar.

The famous double-slit experiment is a cornerstone of this.

Here's the simple version.

You take a coherent light source, like a laser, and shine it at a barrier with two narrow slits. Behind the barrier, you place a detector screen.

When both slits are open, and you don't measure which slit each photon goes through, something strange happens. Light behaves like a wave. On the screen, you don't just see two bright bands; you see an interference pattern—multiple bright and dark stripes, as if waves have gone through both slits and interfered with each other.

Now here's where it gets weird.

When you set up a detector to measure which slit each photon actually passes through—essentially asking, "Which path did you take?"—the interference pattern vanishes. The photons now behave like particles. You get two distinct bands on the screen corresponding to each slit.

Same setup. Same photons.

The only difference is observation.

The act of measuring which slit the photon goes through changes how it behaves. Before observation, it exists in a 'cloud of possibility'—a wavefunction. After observation, it 'collapses' into a specific outcome: left slit or right slit.

At this point, you might be thinking, "What does this have to do with wealth mindset or manifestation?"

This experiment shows that at the deepest levels of physical reality, things don't behave in a fixed, rigid, classical way. They exist as probabilities until there is interaction—observation, measurement, choice.

Now consider your own life.

Before you choose a path, your future is a cloud of possibilities: careers, relationships, financial outcomes. The moment you narrow your focus, decide, and commit energy and attention, you begin collapsing possibilities into reality.

You're not magically bending physics by thinking about a yacht. But your observation and focus do change what becomes real for you.

Just as the act of observation determines the behaviour of photons, your sustained thoughts, intentions, and decisions shape which potential outcomes crystallise in your life.

While we're in the quantum neighbourhood, there's another concept worth mentioning: entanglement.

Quantum entanglement is when two particles become linked such that the state of one instantly correlates with the state of the other, no matter how far apart they are. Change one, and the other updates—instantly—even if they're at opposite ends of the universe.

Einstein didn't like this. He called it "spooky action at a distance."

But experiment after experiment has confirmed that entanglement is real. It's measurable. It's being used in emerging technologies like quantum computing and quantum cryptography.

I'm not saying you and your bank account are quantum-entangled. But the pattern is worth noting.

Systems can be connected in ways that are invisible but very real. Change one part strongly enough, and other parts are forced to adapt.

Your thoughts, emotions, posture, habits, and environment are all entangled in your daily life.

Change your self-talk, and your behaviour shifts. Change your body language, and your emotions begin to realign. Change your environment, and your decisions start to look different.

We treat these domains as separate—'mind', 'body', 'finances', 'relationships'—but they're not. When you shift one seriously, the others eventually have to move.

From the outside, it often looks like a 'lucky break'. From the inside, there was usually a period where someone relentlessly changed how they saw themselves and how they showed up, and that rippled out through everything.

You've probably heard people say, "Raise your vibration," or, "Match the frequency of what you want," and wanted to switch off immediately. It sounds vague. Fluffy.

Let's strip that language down.

At a physical level:

Your brain runs on electrical impulses and chemical signals.

Your heart generates an electromagnetic field that can be measured outside your body.  A few meters outside.

Your nervous system responds to internal and external stimuli and shifts your state.

When you are fearful, angry, or in a scarcity mindset, your body releases specific chemicals—cortisol, adrenaline—and your nervous system shifts into sympathetic activation: fight, flight, or freeze.

When you are calm, confident, optimistic, and focused, a different cocktail dominates—dopamine, serotonin, oxytocin—and your nervous system moves towards parasympathetic balance: rest, repair, strategic thinking.

These states feel like 'vibrations' because they literally are:

Different heart rhythms.

Different breathing patterns.

Different muscle tensions.

Different brainwave patterns.

Now add your external behaviour:

The way you speak.

The choices you make.

What you say yes or no to.

How you respond to pressure or opportunity.

All of that together creates a kind of signature in the world—a pattern of energy and behaviour that other people sense, even if they can't articulate it.

That's your so-called frequency.

So when we say "thoughts become things," we're not saying you can sit on your sofa, chant "Ferrari," and watch one materialise in your living room.

We're saying:

Your repeated thoughts create emotional states.

Those states drive your behaviour.

Your behaviour shapes your outcomes.

If you then use quantum mechanics as a metaphor, you get a powerful picture.

Before observation, a photon exists in a state of potential—many possible paths.

Before decision and commitment, your life exists in a similar state of potential.

By visualising and believing in a desired outcome, you're effectively 'observing' that possibility with intensity. You're choosing it. And then you start taking actions that are coherent with that choice.

That combination—thought, feeling, action—collapses probability into reality.

Now, here's where it all comes together: momentum.

You start with a wealth mindset. You develop faith in yourself and in your future. You practise vivid visualisation and 'living in the end.'

Then something interesting happens.

You wake up with a slightly better attitude. You decide to smile in the mirror, even if it feels forced at first. You stand a little taller. You tackle one small task you'd normally avoid.

Nothing dramatic. Just small changes.

But those small changes send new signals.

Inside your body, chemistry shifts. Stress reduces a notch, clarity improves. To the outside world, your body language, tone, and presence change.

You leave the house, and your energy starts to radiate something new. People on a similar wavelength—driven, optimistic, focused—feel more comfortable around you. Conversations change. Opportunities appear that never seemed to be there before, or that you were too shut down to see.

It's like dominoes falling;

One belief leads to one visualisation.

The visualisation leads to a small shift in emotional state.

The emotional state changes your posture, tone, and choices.

Those changes alter how others respond to you.

Their new responses reinforce your belief.

Your belief deepens your visualisation.

The whole thing snowballs. That's attraction as momentum, not a one-off miracle.

Let's tie this back clearly to the things I have already covered in previous chapters.

In Chapter One, you shifted from "I hope I'll be okay" to "I am building wealth. I am capable. I am on my way." That started wiring new neural pathways, training your brain to expect and handle abundance instead of permanent scarcity.

In Chapter Two, you saw that faith is not just religious language; it's functional. It's the deep conviction that what you move towards can and will meet you halfway, and that you are an active participant—not a spectator—in your financial reality.

In Chapter Three, you stopped treating your dreams as distant wishes and started practising "living in the end". You used your imagination deliberately, rehearsing your future until it felt emotionally familiar.

Now, with this chapter, we've added the science that sits under those shifts:

How your smile, frown, and posture send feedback loops into your brain and into social interactions.

How language and inner dialogue shape your perception and behaviour.

How confidence alters how others rate your competence—and how you rate your own.

How the observer effect gives us a new way to think about focus and choice.

How entanglement is a useful metaphor for the interconnectedness of your thoughts, habits, and results.

How 'vibration' is really just your physiological and behavioural state—and how that attracts or repels people and opportunities.

Reality is not just some mystical fog. It is, in many ways, a measurable interaction between your inner world and your outer experience.

So what do you do with all of this?

This isn't a physics textbook. I'm not asking you to solve equations.

What matters is application. Start noticing your language. Every time you say "I am...", pause and ask: "Is this a command I want to keep sending to my brain?"

Pay attention to your body. Are your shoulders up or down? Is your breathing fast or slow? Are your eyes meeting the world or avoiding it? You can change your state just by changing these.

Use visualisation daily, even for five or ten minutes. See and feel the end result. Get used to it. Make it emotionally familiar.

Watch your momentum. One good decision today can lead to ten better decisions next month. One small shift in posture today can open doors six months from now, because you walk into rooms differently.

The science of attraction is not about hoping the universe drops gifts at your feet. It's about understanding that:

Your mind is a powerful observer.

Your body is a powerful transmitter.

Your repeated choices create momentum.

When you combine a powerful mindset, deep faith in yourself, vivid visualisation, and an understanding of how your inner world shapes your outer actions, you stop waiting for life to happen.

You become, in a very real and scientific sense, the observer that shapes the experiment.

The experiment is your life.

# Chapter 5: Mysteries of Ancient Constructions

**"I don't believe that consciousness is generated by the brain. I believe that the brain is more of a receiver of consciousness."**

**Graham Hancock**

This chapter is very different, and I hope that later on, in the closing chapters, you'll see how the first four chapters and chapters like this start to mutate to create a wider landscape view.

Up to now, we've been talking about wealth, faith, mindset, visualisation, and the science behind attraction. Pretty personal. Pretty 'here and now'.

In this chapter, I'm going to zoom way out—in time and in scope.

Because if you really want to understand how deep some of these ideas go—geometry, vibration, alignment, hidden knowledge—you can't just look at modern psychology books. You have to start looking at stone. Ancient stone. Structures that should not exist, built by people who supposedly had 'primitive' tools, no electricity, no computers, no advanced engineering.

Yet somehow, they left us monuments that we still can't fully explain.

Let me start this chapter with someone you've probably heard of: Graham Hancock.

## Graham Hancock and the 'Forbidden' Past

Graham Hancock has dedicated years to exploring what people like to call 'forbidden history'. He's spent decades digging through ancient maps, declassified scans, obscure academic papers, and so-called 'suppressed' archaeological data.

What he's uncovered has challenged even the most seasoned researchers and, at the very least, raises serious questions about who really built some of the world's most iconic ancient structures—and when.

Hancock is not your typical historian. He's not an Egyptologist. He's not a tenured professor. He's not locked into the traditional academic establishment. In fact, this outsider status is precisely why many people trust him. He started out as a journalist for The Economist—very much part of the mainstream—and then, in the 1990s, he made a major shift: from chasing headlines to chasing deeper truths. What he calls 'lost history'.

His questions are simple but provocative: what if ancient civilisations are far older than mainstream archaeology suggests? What if our history books only reveal half of the story? What if the truth has been concealed beneath centuries of sand, stone, and academic gatekeeping?

For over thirty years, Hancock has travelled the world—from the pyramids of Egypt to the jungle temples of Cambodia and Indonesia, from submerged ruins off the coast of India and Japan to the stepped pyramids of Central and South America. He talks to geologists, astronomers, linguists, mythologists, and engineers. He doesn't stay in one lane. And that's partly why his work hits a nerve: he looks for patterns that specialists, locked in their own fields, might miss.

Plenty of critics dismiss his theories as pseudoscience. Fine. That's the label you give someone who steps outside the approved script. On the other hand, millions of readers and viewers see him as a whistleblower—someone willing to say out loud that a lot of the pieces of our past simply don't fit together.

He's not saying, "Here is the final truth." He's saying, "Look at these anomalies. Look at these numbers. Look at this alignment. Does the official story really cover this?"

Among all the mysteries Hancock digs into, one stands out above the rest: the Great Pyramid of Giza.

## The Great Pyramid: The 'Khufu Problem'

Let's talk about the so-called 'Khufu problem'.

Officially, the Great Pyramid of Giza was built around 2500 BCE, during the reign of Pharaoh Khufu (Cheops). The story goes that it was constructed over just twenty-odd years, using about 2.3 million limestone blocks, moved hundreds of miles with wooden sleds, ramps, ropes, and a huge workforce of paid labourers—not slaves.

You've probably seen the familiar images in schoolbooks and museum displays.

But here's what's strange.

For such a meticulous civilisation as ancient Egypt—which carved its stories, its prayers, its shopping lists, its religious texts into almost every available surface—the interior of the Great Pyramid is eerily silent.

There are no long rows of hieroglyphs, no painted scenes of the afterlife, no detailed inscriptions praising Khufu as the divine builder. If you compare this to other royal tombs in Egypt, which are absolutely packed with text from the Book of the Dead, depictions of gods, and assertions of divine lineage, the Great Pyramid stands out for its lack of detail internally.

Hancock's question is straightforward: if this was really built as Khufu's tomb, why is there no clear internal text declaring that? Why isn't it covered in "I, Khufu, did this" in the same way other pharaonic structures are?

We'll come back to this silence later, because it links into something else entirely: purpose and function.

But first, let me bring in my own perspective—the maths, the sheer mechanics.

## The Maths of Madness: 13 Blocks an Hour

When I started looking at the Great Pyramid, I couldn't get away from the numbers. Not the mystical ones at first. Just the practical ones.

Renovating my own house in France, I know how long it takes to position one tile or one block properly. And those are regular household-sized blocks, not two- to fifteen-ton limestone monsters.

Let's just walk through the basic arithmetic that the history books casually brush over.

We're told that around 2.3 million blocks were used, and the build time was roughly twenty years. Let's be generous and assume they worked every single day, no days off, no holidays, no breaks for floods, famine, disease, politics, or "my back's gone".

Twenty years is about 7,300 days. Divide 2.3 million blocks by 7,300 days, and you get roughly 315 blocks per day. Now divide that by 24 hours (again, being generous—assuming non-stop work, day and night), and you arrive at about 13 blocks per hour.

In other words, on average, they had to quarry, shape, transport, lift, and precisely place one of those huge stone blocks roughly every four to five minutes. For twenty years. Non-stop. No mistakes. No, "we mis-cut this one, lads, start again."

Remember, these blocks range in size, but some of them are the size of a large coffin—except made of solid limestone or granite, weighing multiple tons. If you've ever tried to lay one large concrete lintel or manoeuvre a steel beam with machinery, you'll know how absurd this is.

This chapter is not about giving a definitive answer to how the pyramids were built; entire books do nothing but argue about that. What I'm saying is simple: the standard story we were all fed in school is, at best, oversimplified. At worst, it's covering up something we still don't properly understand.

And that's where people like Christopher Dunn and Scott Creighton come into the picture.

## Christopher Dunn and his book 'The Giza Power Plant'

Christopher Dunn is not an archaeologist. He's a master engineer and machinist. He's spent decades working with precision tools, manufacturing, and industrial processes.

In his book 'The Giza Power Plant', Dunn approaches the Great Pyramid like an engineer, not a priest. Instead of asking, "What does this symbol mean?" he asks, "What does this thing do?"

He points out some interesting clues. First, the precision: some surfaces inside Egyptian structures—like the sarcophagus in the King's Chamber—show machining marks and levels of flatness that are comparable to modern machine tools. We're talking tolerances that don't fit the picture of 'men with copper chisels banging away blindly'.

Then there's the geometry and alignment: the Great Pyramid is aligned to true north with astonishing accuracy. Its dimensions embed key mathematical constants. The ratio of the pyramid's height to its base perimeter links to π (pi). The slope angle and internal proportions echo other fundamental ratios, including ones linked to the Earth itself. This is also touched on in a later chapter.

Even the choice of materials is telling. Different stones are used strategically—limestone, granite, sometimes basalt—each with different physical and acoustic properties.

Dunn's central hypothesis is controversial, but fascinating. He believes the Great Pyramid may have functioned, in some way, as a gigantic machine—specifically, as a kind of power plant or resonant energy device harnessing vibration.

In other words, not just a tomb. Not just a big pointy pile of rocks. But a tuned system.

Whether you agree or not, it changes the way you look at the structure. You stop seeing it as dead stone and start seeing it like an engineered device.

And this leads us straight into another challenge to the traditional narrative: who really put their mark on this thing?

## Scott Creighton and 'The Great Pyramid Hoax'

Scott Creighton's book 'The Great Pyramid Hoax: The Conspiracy to Conceal the True History of Ancient Egypt' focuses on a specific part of the story: the inscriptions that supposedly 'prove' the Great Pyramid belonged to Khufu.

In the 19th century, an explorer named Colonel Howard Vyse claimed to have discovered painted quarry marks and cartouches in hidden chambers above the King's Chamber, including one that bore the name Khufu. These inscriptions have since been taken as the smoking gun that settles the debate: 'Case closed, it's Khufu's pyramid.'

Creighton re-examines this story and raises a very uncomfortable possibility: that some of these marks might have been faked or misrepresented to secure fame, money, or recognition at the time. He questions the circumstances of the 'discovery', highlights inconsistencies in the hieroglyphs, and points out the lack of other clear, contemporary references inside the structure.

Now, whether you think Creighton is right or not is almost secondary here.

The crucial point is this: if the key evidence linking Khufu to the Great Pyramid is even potentially compromised, the entire official story starts to wobble.

Then we're left with a giant, hyper-accurate, mathematically advanced, acoustically tuned, megalithic structure… with no clear builder's signature and no clear function written on the label.

That should make anyone who cares about logic and evidence raise an eyebrow.

## Sound, Vibration, and F-Sharp: A Tuned Stone Box?

Out of all the rabbit holes I've gone down, one of the most mind-bending is the vibrational aspect of the Great Pyramid. And I cover this in some more detail in the next chapter.

The Great Pyramid's mathematical proportions and internal layout are thought by some researchers to give it unique acoustic properties. The exact shape and dimensions of chambers and passageways may allow certain frequencies to resonate and amplify within.

In particular, some studies and experimental work have suggested that parts of the Great Pyramid respond strongly to frequencies around the musical note F-sharp (around 370 Hz). Not just as a vague hum, but through resonance, standing wave patterns, and amplification.

Is that a coincidence?

Maybe. Or maybe the ancient builders understood sound and vibration far better than we give them credit for, and they built a structure that could harness, amplify, or interact with specific frequencies—whether for ritual, healing, communication, or something more technical.

This idea doesn't exist in isolation, either.

Think of Tibetan singing bowls, used in meditation and healing, tuned to particular frequencies that people report as having distinct emotional and physical effects. Think of chanting and drumming in Native American and other indigenous traditions, where repetitive rhythms and tones are used to shift consciousness, connect spiritually, and heal. Even in Western music, composers have long used key signatures to evoke specific moods. Modern songs like Beyoncé's 'Halo' in F-sharp major or Taylor Swift's 'All Too Well' in F-sharp major are built around emotional resonance as much as melody. Different keys feel different because they interact with our nervous system in specific ways.

We'll dig more into frequencies in the next chapter, but for now, think about this: if a modern neurologist can stick someone in an MRI machine and show that certain sounds and frequencies change brain activity, and an ancient builder spends an absurd amount of effort to create a stone structure that just happens to resonate at very specific tones, maybe they were not idiots piling rocks for no reason.

Maybe they were manipulating energy and consciousness with architecture.

There's another detail that often gets brushed past.

### No Soot, No Smoke, No Simple Story

Inside the Great Pyramid, and in many other deeply enclosed structures, there's a noticeable lack of soot or burn marks you'd expect if people had been working extensively with open flames for lighting.

Mainstream answers have been offered—burnished reflectors, limited torch use, and so on—but when you combine tight, complex passageways with massive, precisely cut stone blocks in deep interior spaces, and yet find no obvious scorch marks or heavy soot layers, it raises another simple question: how exactly were they illuminating these spaces to work?

Again, it's not definitive proof of some lost technology, but it's another data point in the growing pile of 'things that don't quite add up'.

Alignments with stars. Geometric patterns. Mathematics embedded in stone. Advanced acoustics. Mysterious cleanliness, electricity. The list goes on. And it doesn't stop in Egypt.

## Beyond Egypt: A Global Pattern of Impossible Stone

If this were all limited to the pyramids at Giza, you might be able to write it off as an anomaly.

But it isn't. The same style of 'impossible' stonework and baffling alignment shows up all over the planet—in cultures that, on paper, were not supposed to be connected.

Let's take a brief tour.

The Sumerians, in ancient Mesopotamia (modern-day Iraq and surrounding areas), are often credited with some of the earliest known writing, advanced city-states, and surprisingly complex astronomical and mathematical systems. They were mapping the movements of planets and stars thousands of years ago. Their myths speak of knowledge passed down from earlier 'gods' or civilisers. Their base - 60 number system still echoes in our 60 - second/minute and 360 - degree circles.

We're told they emerged out of nowhere—suddenly advanced, with writing, law codes, agriculture, and architecture. Where did that leap come from? How did a supposedly 'primitive' society suddenly have detailed astronomical records and complex formal systems? It's another early example of what looks like a jump rather than a slow, linear crawl.

Most people don't realise that China has its own pyramids—large, ancient, often overgrown mounds and stepped structures, some of which have only recently started to be studied properly. Some of these structures show astronomical alignment. Others are linked to imperial tomb complexes shrouded in mystery. There are written records of palaces filled with rivers of mercury and mechanical traps. China also has very old temples and megalithic structures that combine sacred geometry, cosmology, and ritual use of sound. For generations, some of these sites have been kept relatively quiet or restricted—either for political reasons or simply because they didn't fit nicely into the standard global story. Again: advanced geometry, cosmic alignment, stone, and mystery.

Travel across the world to South America, and you hit structures that are just as mind-bending as anything in Egypt. Places like Sacsayhuamán near Cusco in Peru, Ollantaytambo, and other sites are attributed to the Incas and even pre-Incan cultures. Here you see enormous stone blocks—some weighing tens of tons—fitted together with such precision that you can't slide a piece of paper between them. The stones have multiple sides, weird angles, and rounded edges. Some look like they were softened and moulded into place rather than carved with simple chisels. The joins are tight, interlocking, and earthquake-resistant.

Ask a modern stonemason what it would take to produce joints like that—on that scale, at that altitude, with no modern cranes—and see what they say.

On top of that, some of these sites show intriguing magnetic and energetic anomalies. Compasses misbehave. People report unusual sensations. Again: vibration, geometry, and stone.

You can continue this world tour and keep hitting the same theme. Baalbek in Lebanon is home to some of the largest hewn stones ever placed by humans—single blocks weighing hundreds of tons, stacked with ridiculous precision. Göbekli Tepe in Turkey is a massive stone temple complex pre-dating Stonehenge by thousands of years, with intricate carvings and advanced organisation that completely rewrites the idea that 'hunter-gatherers were simple. Across Europe, megalithic circles and alignments—from Stonehenge to lesser-known rings and rows—line up with solstices, equinoxes, and star rises.

The consistent pattern is this: ancient people aligning stone with the sky, embedding maths and geometry in construction, working with vibration and acoustic resonance, whether in chambers or open spaces, and leaving behind puzzles that our 'advanced' civilisation still argues about.

It starts to look less like a random myth and more like a long, interrupted tradition of using architecture as technology.

## Why Does Any of This Matter for You?

You might be thinking: "Okay, Lewis, this is all interesting, but what does this have to do with me building wealth, shifting my mindset, or getting an investment for my start-up?"

Fair question.

Here's why I'm including this in a book about mindset, wealth, and attraction.

Firstly, patterns repeat. Just as Fibonacci shows up in shells, charts, buildings, and art, the same deeper patterns—geometry, resonance, belief, alignment—show up in how human beings have tried to interact with reality for thousands of years.

Secondly, knowledge can be hidden or lost. Just as there are rumours of secret rooms in the Vatican and lost scrolls in the Library of Alexandria, there are also 'lost' or buried techniques in our own lives—ways of thinking and using our minds that we've forgotten or dismissed. If entire civilisations can have their advanced capabilities wiped from the official record, don't you think it's possible that modern individuals can have their own advanced inner capabilities buried under fear, doubt, and conditioning?

Third, function is often wrapped in a story. If a pyramid can be a machine disguised as a tomb, then a belief system can be a technology disguised as religion or myth. The stories about gods, giants, and great builders may be one level of description. The geometry, sound, and alignment are different. Likewise, your 'story' about yourself is the surface layer. Underneath, your thought patterns, visualisations, and emotional states are the real machinery.

Finally, momentum and resonance aren't new ideas. These ancient structures may literally be physical embodiments of resonance and momentum: energy in, amplified output. That's not far off what we're doing mentally and emotionally with visualisation and belief.

This chapter isn't here to turn you into an amateur archaeologist. It's here to widen your lens.

When you realise that ancient builders may have been using geometry and vibration deliberately—and that their work still puzzles modern engineers—you stop casually dismissing ideas like 'your thoughts have power', 'your words carry frequency', and 'your belief structures shape your reality'.

You start seeing that, on some level, human beings have always tried to interface with reality through geometry, sound, intention, and alignment.

The Great Pyramid, the stone walls of Peru, the Sumerian tablets, the hidden Chinese pyramids—they are all physical reminders that there is more to this place than we're told in school.

## Closing the Loop (For Now)

Let's circle back to where we started: a journalist like Hancock questioning the official story; an engineer like Dunn treating the pyramid as a machine, not a tomb; a researcher like Creighton questioning the authenticity of key 'proof'; and global sites that quietly defy our understanding of 'primitive' cultures, with acoustic and vibrational properties built into stone.

And my own simple, stubborn question: what if all of this isn't random?

What if, just as we use coding languages and circuit boards to manipulate electrons and information today, ancient builders used stone, shape, sound, and orientation to manipulate energy and consciousness?

In the first four chapters, we looked at how your mindset sets the foundation; how faith and belief converge with fortune; how visualisation allows you to live in the end result; and how science—from psychology to quantum mechanics—shows that your inner world shapes your outer experience.

Now, with this chapter, we've stepped back into time to look at how earlier civilisations might have been doing something very similar—just on a massive, physical scale.

You don't need to accept every theory about ancient power plants or secret societies to benefit from this. You just need to see the pattern: hidden knowledge, geometry, vibration, alignment, intention, momentum.

I'm going to explain some more about these themes later.

Because once you start seeing the world—past and present—as an interconnected system of patterns and resonance, you stop playing small with your own potential.

You begin to realise that your mind, your emotions, and your beliefs are not soft, fluffy things. They are tools. Technologies. And just like the builders of the Great Pyramid, the Incas, the Sumerians, and the ancients of China, you can learn to use them with precision.

We're not done with frequencies and resonance yet. In the next chapter, we'll go deeper into exactly why specific tones, patterns, and alignments—whether in stone or in your nervous system—are far more important than most people ever realise.

# Chapter 6: Frequencies

**"If you want to find the secrets of the universe, think in terms of energy, frequency and vibration."**

**Nikola Tesla**

This chapter isn't about chimes and chants. It's a very important piece of the puzzle.

Roughly 99.9% of everything is made of empty space. The phone in your pocket, the floor under your feet—even you. At the atomic level, what we call 'solid' is mostly space, with particles vibrating at different frequencies.

You are, in essence, frequency.

So is everything else in the entire universe.

When you change your internal state, the outer world has to reorganise to match that state.

You don't simply 'think thoughts' in isolation—you tune into them. When you tune into fear, you receive only negative outcomes. You've probably had the experience: bills are piling up, money is scarce, and then your car breaks down. As if the lack of money wasn't enough, you now have an extra cost you didn't anticipate.

To get to work, you need a car. To pay for the car, you need to work. A psychological chokehold. You feel trapped in a vicious cycle.

What happens next? You lose faith. The visualisations no longer feel 'real'. You slip back into old habits and old ways of thinking.

But remember what I covered in Chapter 4: attraction also works with positive sentiment. Momentum. Or is it something else? What if you think of it as matching frequencies?

Frequencies attract frequencies.

When I started digging into this, I found information on several sound frequencies that are said to have 'special' significance. If you ask Alexa or Google Home to play music at these frequencies, or if you search for them on YouTube, you'll find entire playlists built around them:

- 528 Hz is sometimes called the 'Love Frequency', believed to promote healing and transformation, enhancing feelings of love, clarity and peace.

- 396 Hz is said to help alleviate fear and guilt, contributing to emotional clarity and a more positive mindset.

- 417 Hz is associated with facilitating change and 'undoing' situations, helping to clear emotional blockages and restore balance.

- 432 Hz is thought to resonate with the universe and nature, promoting harmony and emotional stability.

- 639 Hz is linked to improving communication and relationships, deepening connection and emotional clarity in interactions.

Whether or not you buy every claim made about these specific numbers, the broader point stands: sound frequency affects the state of mind.

As I listened and researched more, I stumbled into another area: how we develop as humans, and how the mind of a child can learn languages or complex skills so quickly. This led me into brainwaves—especially theta waves—and how they relate to learning, habit change and even hypnosis.

Brainwaves are usually grouped like this:

- Delta (0.5–4 Hz): deep sleep and physical repair.

- Theta (4–8 Hz): creativity, intuition, deep meditation, and that 'in-between' drifting state.

- Alpha (8–14 Hz): relaxed focus, reduced stress, a kind of calm alertness.

- Beta (14–30 Hz): active thinking, concentration, everyday problem-solving.

- Gamma (30–100 Hz): high mental activity, complex processing, peak performance.

Now imagine you could deliberately use sound to shift your brain into different states.

Imagine 'tricking' your brain into a theta-like state—the same kind of state you moved through more naturally as a child—so you could re-open that learning window. Delete bad habits. Re-programme yourself for a wealth mindset.

Think tuning forks!

Guess what else, hypnosis is strongly associated with theta. That explains a lot.

And hypnosis isn't new. Ancient cultures were inducing altered states of consciousness long before anyone used that word. The Egyptians, for example, used rituals, chanting and music designed to put people into trance-like states conducive to healing or spiritual experience. They had 'sleep temples' where individuals would rest, dream and enter altered states as part of a healing process. In modern language, they were experimenting with subconscious access.

We've already covered some of the physics, but it's worth briefly returning there.

In quantum mechanics, particles such as electrons and photons exhibit both particle-like and wave-like behaviour. Louis de Broglie proposed that every particle can be associated with a wavelength. Waves are described by amplitude, frequency, wavelength and velocity. Sound waves are longitudinal waves—vibrations travelling through a medium. When you analyse sound in the frequency domain, you can decompose a complex sound into its constituent frequencies to see what's really there.

The Egyptians may not have used words like 'wave-particle duality', but they clearly understood, at a practical level, the effects of sound. In rituals and ceremonies, they used music and chanting to influence emotions, shift states and create a sense of connection to the divine. They seemed to grasp resonance and vibration as tools.

And, as with so much else, it's very likely some of that knowledge was reserved for a select few—priests, initiates, scholars. That secrecy creates a mystique around it and, over time, the perception that the knowledge was 'hidden'.

My question is: why?

By now, you've probably noticed a pattern.

We started with mindset, faith, visualisation, and attraction. Then we jumped into ancient structures, pyramids and geometry. On the surface, those topics seem miles apart. But when you zoom out, they're not random at all. They all point toward the same thing:

Frequency.

Energy.

Information.

And how humans have tried to interact with them throughout history.

For most people, Pythagoras is just 'that triangles guy' from school. But in his own time, Pythagoras was obsessed with something far stranger: the music of the spheres.

He discovered that musical intervals—octave, fifth, fourth—follow exact mathematical proportions: 2:1, 3:2, 4:3. Simple ratios. Clean numbers. If you tighten or loosen a string to those precise proportions, the sound 'locks in.' It feels right. It resonates.

If these same proportions govern harmony, maybe they also govern the movement of the planets. Maybe the entire cosmos is a kind of symphony, each orbit a note in an invisible score. You can call that poetic if you like, but physics later discovered orbital resonances and gravitational harmonics. Once again, the poet-mystic got there before the lab coat.

Let's look at one more piece of that puzzle: crystals.

Not as jewellery. Not as 'nice shiny rocks' on a shelf. But as resonant structures, information carriers and tools used by both ancient cultures and modern science in ways that look strangely similar once you strip away the labels.

Some people hear the word 'crystals' and instantly switch off. They picture incense, vague affirmations, and someone claiming a rock will solve all their problems.

That's not where I'm going with this.

I'm interested in why shamans, priests and temple builders from different cultures valued crystals so much—and why, in the 21st century, some of our most advanced technologies are quietly doing the same thing.

## Crystals: Order, Pattern, Frequency

A crystal is not 'magic' in a fairy-tale sense. It's a specific kind of material structure.

In a crystal, atoms or molecules are arranged in a repeating, highly ordered pattern. That internal order creates predictable physical properties, very stable vibrational behaviours and precise ways of interacting with light, electricity and pressure.

Different crystals—quartz, amethyst, sapphire, tourmaline—have different internal structures and, therefore, different properties. But the key idea is simple:

crystals are nature's way of organising matter into stable, repeating patterns.

That alone should ring a bell, given everything we've already discussed: frequency, vibration, resonance, patterns that repeat across scales.

Now compare two apparently different worlds:

- Ancient shamans and temple cultures.

- Modern engineering and data storage.

Are we really as far removed from them as we like to think?

Across the world, in cultures that supposedly had no contact with each other, you find a recurring theme: crystals used in ritual, healing, communication and spiritual work.

Indigenous shamans in places like South and North America, or Siberia, often used quartz crystals in ceremonies. These weren't fashion accessories; they were tools for focusing intention, 'calling in spirits' or amplifying energy. Crystals were 'fed' with breath, smoke, and intention.

The ancient Egyptians used lapis lazuli, carnelian, turquoise and clear quartz in jewellery, burial items, temple constructions and priestly tools, all with specific symbolic and energetic meanings. Chinese traditions prized jade and certain stones for their connection to qi—longevity, protection, harmony. Hindu and Buddhist systems used crystals and gemstones in malas, statues and ritual objects, each linked to certain energies or mental states. Mesoamerican cultures carved obsidian mirrors and crystal objects used in divination and ritual; mirrors were seen as portals, not just reflective glass.

From a modern, cynical viewpoint, it's easy to write all that off as superstition.

For me, too, a lot of this I am on the fence with. However, there is too much in terms of coincidence for me to simply ignore.

But then you have to ask: why did so many different cultures, across different eras, go to such lengths to mine, shape and polish specific stones, and embed entire belief systems into them? Why not just use any rock, or wood, or clay?

Either they were all collectively insane... or they noticed something, tapped into something that has since been forgotten or hidden.

## Frequency, Structure, and Amplification

Because of their ordered structure, crystals have predictable vibrational properties. Expose them to certain stimuli—pressure, heat, electrical fields—and they respond in consistent ways.

Quartz is the classic example.

Quartz has a property called piezoelectricity. In plain language:

- Press or stress a quartz crystal and it generates an electric charge.

- Apply an electric field, and the crystal physically deforms (very slightly) and vibrates.

This isn't theory; it's engineering. Quartz is used in watches to keep accurate time because it vibrates at a highly stable frequency. It appears in microphones, speakers, sensors, transducers, and timing circuits.

Quartz isn't just sitting there looking pretty. It's doing something. It responds to energy and produces energy. It converts between mechanical and electrical signals. It vibrates at specific, stable frequencies that can be used as reference points.

Now imagine you're an ancient culture without a lab coat or oscilloscope.

You don't know the phrase "piezoelectric effect", but you've observed that certain stones behave differently. You've noticed particular crystals feel 'powerful' in ritual. You've seen that placing crystals in certain arrangements seems to heighten experiences—emotional, spiritual, and cognitive.

You might not talk about volts or hertz, but you'd have your own language: amplifiers, protectors, bridges between worlds, carriers of the gods' energy.

Strip away the mythology and what's left? Humans experimented with frequency and structure using the tools they had: stone, sound, light, repetitive ritual.

You do the same thing today with your phone and Wi-Fi. You just call it 'tech' instead of 'ceremony'.

## Crystals as Information Devices

Here's where it gets even more interesting.

Modern science is now doing with crystals what ancient people hinted at metaphorically: using them to store information.

There's ongoing research on using quartz glass to store data in what's sometimes marketed as '5D optical data storage'. The basic idea:

- Ultra-fast laser pulses are used to write tiny patterns inside quartz glass.

- Those nano-structures change how light passes through—altering polarisation and other optical properties.

- Each tiny point can encode multiple bits of information.

- The result is a storage medium that could potentially last for billions of years without degrading under normal conditions.

In other words, you can take information, encode it into structured light, burn that pattern into quartz, and read it back later.

We're using crystals as long-term memory devices. Stable, high-density storage. Physical matter holding patterns that outlast the civilisation that created them.

Is that really so different from the idea that ancient cultures might have seen crystals as 'record keepers' or 'carriers of wisdom'?

Again, the language has changed. The underlying behaviour hasn't. We now openly acknowledge that crystals can hold data, respond to and transform energy, and act as stable frequency references.

So, when an ancient shaman says a crystal 'holds memories' or 'carries messages', is that pure fantasy? Or is it a poetic description of something we're now rediscovering in clinical language?

## Why Healers Use Crystals (and What That Means for You)

Step into the mindset of an ancient healer.

A human being is not just flesh and bone; it's an energy system: breath, emotion, thought, spirit. A field.

When someone is sick or traumatised, their energy is seen as disturbed, blocked, or fragmented.

If you're working at the level of energy and frequency, what kinds of tools would you be drawn to? Things that vibrate predictably. Things that calm, focus or direct attention. Things that react to pressure, heat, and touch.

Crystals tick all three.

So you place them on the body at specific points, in circles around a person, on an altar at the centre of a ritual space, or into amulets worn close to the skin. The logic is simple:

'If this stone has a stable, coherent pattern, and this human has a chaotic, disrupted pattern, maybe bringing them into contact will help tune or align the human.'

From a scientific standpoint, sometimes those effects are measurable; sometimes they're subtle. From the healer's standpoint, results matter more than journals. People feel different; they behave differently; they heal. So the practice continues.

Now connect this back to everything we've been talking about.

In mindset work, you're trying to stabilise your mental signal. In visualisation, you're trying to consistently resonate with a desired future. In wealth building, you're trying to maintain a stable internal state while the market and the world around you swing violently.

Crystals, in this view, are physical metaphors and tools for coherence. Training wheels for your frequency.

## The Bridge Between Inner and Outer

A core theme of this book is that your inner world and outer world are not truly separate.

Thoughts become chemicals.

Emotions become posture.

Posture becomes behaviour.

Behaviour becomes results.

Crystals sit in an interesting place in that chain.

On one hand, they are physical objects with measurable properties. On the other, they've been used for thousands of years as tools for working with intangible things: intent, prayer, consciousness, 'spirit'.

Imagine holding a crystal in your hand and focusing on a specific intention—wealth, healing, clarity. Several things happen at once:

You focus. Instead of your mind flitting around, your attention anchors to a single point. That alone changes your brain state.

You build an association. Over time, the crystal becomes linked to a particular mental and emotional pattern. Touching it later can trigger that state—like an anchor in NLP.

You shift physiology. If this practice calms you, your breathing slows, your muscles relax, your face softens. Your nervous system moves out of fight-or-flight and into a more creative, strategic mode.

And then others respond differently to the calmer, clearer version of you. New feedback loops start.

Is the crystal doing this by itself? No. You are. The crystal is a symbol and tool—a piece of 'technology' you're using to modulate your own frequency and focus.

But even on a strict material level, the crystal interacts with fields—electromagnetic, thermal, and mechanical. The exchange might be subtle, but it's real.

So it's not completely wrong to talk about 'energy' here. It's just that a lot of the magic is happening through your own nervous system and attention.

## Crystals, Memory, and You

Back to data storage.

If we can encode information into quartz so that it lasts longer than our current civilisation, it suggests a few things: information can be embedded in patterns at deep levels; it doesn't need to look like a USB stick to be a memory device; once encoded, it simply sits there until someone with the right tools reads it.

Now run a thought experiment.

What if some ancient structures or ritual objects were doing something similar—not with binary files, but with patterned energy?

Crystals embedded in walls, altars, staffs, crowns, jewellery. Arranged in specific formations. Aligned with stars. Activated with sound—chants, drums, singing—and repeated ritual action.

You have structure (the crystal lattice), input (sound, breath, touch, intention), and output (shifts in human consciousness and behaviour).

Is that so far removed from quartz glass, laser pulses and reading data back out?

Different tools. Same principle.

Here's where it becomes personal: your brain and body are also storage devices.

Every repeated thought pattern is like a file you've saved. Every emotional loop is a program that runs by default. Your posture, tone and language are the user interface.

In that sense, crystals and other physical tools are 'external memory holders' or 'state triggers' that can help you rewrite your own internal files.

When you pair them with a wealth mindset, faith in yourself, visualisation, and an understanding of frequency and momentum, they become part of a toolkit, not a superstition.

## Why This Matters for Wealth and Success

All of this comes back to how you interact with reality.

Ancient cultures used crystals to focus attention, align group rituals, anchor stories and beliefs, and work with unseen forces.

Modern science uses crystals for stable timing, precise measurement, data storage, frequency generation and filtering.

You are using your mind and emotions to drive decisions, attract or repel opportunities, build or destroy momentum, and stay stable under pressure.

The common denominator is simple:

Frequency + Structure = Function.

The structure of a crystal determines how it responds to frequency. The structure of your beliefs determines how you respond to life. The structure of your daily habits determines the 'signal' you live in.

Crystals remind us that matter can hold patterns, patterns can be stable over time, and stable patterns can be used as reference points in a chaotic environment.

When you choose a belief like "I am capable of building wealth" and repeat it with conviction, you're trying to crystallise that pattern inside yourself. When you visualise your future repeatedly and pair it with strong emotion, you're burning that pattern in—like a laser writing data into quartz.

When you surround yourself with physical cues—a crystal on your desk, images on your wall, specific environments—you're creating an external field that reinforces your internal signal.

Is a crystal required for success? No. Let's be clear.

But understanding how crystals work—scientifically and symbolically—gives you another way of thinking about stability, resonance and long-term memory, in you and in matter. And that deepens your understanding of how to program yourself deliberately instead of letting the world do it randomly.

## Pulling It Together

Here's where we are:

- Ancient cultures treated crystals as sacred, powerful and important.

- Modern science treats crystals as precise, stable, high-functioning components—and is beginning to use them as ultra-long-term data storage.

- Crystals are structured matter that responds predictably to frequency.

- Shamans and healers use them to focus intention and shift energy states.

- You can apply the same logic—structure plus frequency—to your own mind, emotions and behaviours.

Crystals are not a shortcut past the work. They won't drop millions into your account while you sleep.

But they are a tangible reminder that order matters, pattern matters, repetition stores, and frequency—whether in stone, sound, thought or feeling—shapes reality.

In the same way that ancient builders encoded knowledge into stone structures that still speak to us, you are encoding your future into your nervous system with every thought, every visualisation, every emotional loop you reinforce.

Crystals are one of the oldest metaphors—and tools—we have for understanding how stable, precise patterns can exist in a chaotic world.

If you're serious about building wealth, freedom and alignment, that's exactly what you're trying to become:

A stable, coherent signal in the middle of noise.

### Wider Lense

At its core, the universe is not made of 'stuff' in the way we were taught at school. What we call matter is just organised energy—oscillating fields nested inside other oscillating fields, from galaxies down to atoms and subatomic particles. There is no solid brick at the bottom; there are only different densities and frequencies of vibration. The human being—brain, body, 'soul', the lot—is built on the same principle: an extraordinarily complex system of energy fields tuned to a very narrow band of reality.

Your brain works like a radio: it locks onto a specific frequency range and filters everything else out, converting universal energy into the colours, sounds, tastes and textures you call 'the world'. Change the tuning and you don't suddenly create new stations; you simply become aware of what was already there.

According to the CIA's Gateway report (more on this in chapter 15), this entire 'world' is a shared hologram: a three-dimensional interference pattern of energy and information. Physicists like David Bohm and neuroscientists like Karl Pribram took this seriously enough to argue that both the universe and the human mind are holographic—patterns within patterns, with consciousness as the interface between the personal hologram (your mind) and the universal one (everything).

The Gateway process is, essentially, a way of hacking that interface on purpose. Robert Monroe (more info on Robert Monroe to come in Chapter 11) discovered in the 1950s that specific sound patterns could push the brain into predictable states: sleepy, alert, anxious, calm. He then went further. By feeding slightly different tones into each ear and letting the brain 'follow' the difference between them—a method later called Hemi-Sync—he found he could synchronise the left and right hemispheres, shifting people out of normal beta-wave, problem-solving mode into deeper, slower, more coherent states of consciousness. In the Gateway model, your usual mental state is like a lamp—light going everywhere, half-wasted. Hemi-Sync turns that same power into a laser: focused, coherent, able to interact with subtler layers of the universal energy field. Once you grasp that consciousness sits at the source of this hologram, the implications are obvious: by changing how you hold and project thought—what the report calls 'patterning'—you are not just having private daydreams; you are altering the interference pattern you live in. In plain English: if reality is a hologram generated from information, and consciousness is the projector, then your repeated thoughts and emotional states are not soft, harmless things. They are commands. That's where ideas like 'The Secret' / 'Law of Attraction' come from—stripped of the fluff, they are crude versions of the same claim: if you can sharpen, intensify and sustain a pattern in your mind, the hologram around you has to adjust.

# Chapter 7: Technology, Religion, & Numbers

**"Mathematics is the language with which God has written the universe."**

**Galileo Galilei**

Technology has a very personal connection with me.

As an inventor with patents to my name, I actually 'fell' into the technology space almost by accident. But once I was in, something shifted. After developing my first device, I found myself with a much deeper understanding of the inner workings—not just what a circuit board did, but why it was laid out the way it was. Why a capacitor and a resistor sat where they did. Why some boards could be a simple single layer and others need four layers or more.

And, much like the chapters before this, I started to feel that what we call 'technology' is less some cold, separate thing and more a momentum phenomenon—a continuation of patterns that have been here all along.

As my knowledge grew, so did my ability to see what was possible.

One night, I had this dream. I saw a product in its entirety. Not a vague idea. The full thing. I knew exactly how it would look, what it would do, who I needed to build it, and how we were going to bring it to life. It was like someone had dropped a finished blueprint into my head.

Years later, I read that Nikola Tesla had visions of devices come to him in almost the same way. He described ideas appearing as vivid images above his head; he could rotate them, inspect them, and 'reach up into the ether and grab them.'

At the time I first read this, I just thought, "That's interesting." But as my research evolved, and more pieces started to overlap—mindset, faith, visualisation, ancient structures, vibration—the more I realised the significance.

Tesla had a boss at one point: Thomas Edison. Later a colleague. Eventually, a rival.

Edison vs. Tesla became the poster child for AC vs. DC current. That story alone is a whole book in itself and a fascinating part of history: Edison betting the house on DC, Tesla seeing the bigger pattern—AC, long-distance transmission, efficiency—and the 'war of the currents' that shaped the entire modern world.

But step away from the personalities for a second and look at the pattern.

At the technology level, we can break almost any product down into two things: an idea, and a pattern of numbers.

Everything started as an idea. This book. The paper or the screen you're reading it on. The first spear. The first wheel. The first computer.

Then, when we move beyond imagination into the physical world, every item—including us humans, our phones, laptops, cars, satellites, MRI scanners—is just organised particles. Structured matter.

Can you see how crazy this can get?

This doesn't mean you can imagine a tree that grows money, and one turns up from Amazon tomorrow. We've covered that. Fantasy without action does nothing.

But what if, like with wave principles in physics—or Elliott Wave in trading—we can break down big, overwhelming things into understandable slices?

That makes innovation easier to comprehend. You stop looking at a finished skyscraper and saying, "That's impossible," and you start breaking it down: one beam, one bolt, one calculation, one layer at a time.

What if the default belief was "anything is possible... if we can find the right pattern" instead of the usual "you can't do that"?

If Edison had completely crushed Tesla's AC system, we'd probably have DC power stations every kilometre or two, burning energy, pumping heat, bleeding efficiency. The grid would be smaller, uglier, and far less scalable.

But one man's conviction—backed by maths and engineering—changed that.

Now, let's bring this back to my own experience.

## Circuit Boards, Running Tracks, and IF–THEN Logic

When I was having my first product developed, I was fascinated by the process.

I already had my own manufacturing company. I'd been building other people's circuit boards: soldering, testing, shipping. But when I commissioned a feasibility study from the local university for my own idea, I suddenly found myself on the other side of the table.

Now I was the inventor.

I didn't just want the end result. I wanted to understand how the thing was going to work. How the developers were thinking.

One of the engineers explained it to me in a way that clicked instantly.

They used a simple analogy. They talked about using chipsets on a breadboard—a test board. A mock-up using parts from different suppliers. Imagine a memory chip from one company and a memory card slot from another. On paper, they 'do the same job', but the internal design, the capacitors and resistors, the timing tolerances—everything—is slightly different.

"Imagine a circuit board like a running track," they said. "The electricity—those tiny electrons—have to get around the track in a certain amount of time. If it's too fast, it doesn't work properly. If it's too slow, it doesn't work properly either. So we add hurdles or lengthen the track using resistors and capacitors. We're effectively shaping time with components."

That landed immediately.

Then you've got the software on top.

At the most basic level, software is just ZERO and ONE. Off and on. True or false. In code, you're basically saying: IF this happens, THEN do that. ELSE do something else. Nested IF–THEN–ELSE statements, millions of them, billions at scale.

Zoom out, and you realise: pretty much anything is possible—as long as you can express it in logic and numbers.

In my mind, the simplified logic was this:

If I can visualise it clearly enough, and someone can translate that visual into numbers, components and IF–THEN rules, then it exists in potential. And if I believe in myself enough to keep pushing, it can move from potential to physical reality.

The logic? It's all math.

Keep in mind: I teach trading. Math is everywhere.

## Mandelbrot, Fibonacci, and the Language of Reality

You may or may not have heard of the Mandelbrot set.

On paper, it's 'a set of points in the field of complex dynamics.' In reality, once you see it, that bland description falls apart.

Named after Benoit Mandelbrot, it's a fractal: an infinitely complex pattern generated by a simple formula repeated over and over.

Zoom in, and you see endless spirals, self-similar shapes, beauty and symmetry at every scale. Change your zoom, and the pattern still looks like itself.

The Mandelbrot set is a visual doorway into a core idea:

Simple rules + repetition = infinite complexity.

Nature does this constantly.

As a trader, this collides with another rabbit hole: Fibonacci.

The Fibonacci sequence is simple:

0, 1, 1, 2, 3, 5, 8, 13, 21...

Each number is the sum of the previous two. That's it.

From that simplicity, you get ratios like 61.8% that show up everywhere: in spiral galaxies, sunflower seed patterns, hurricanes, architecture, art, and even in the proportions of your own body.

And yes, in stock charts.

Price swings often retrace around 38.2%, 50%, 61.8%. Not perfectly, but often enough that ignoring it is just bad business.

Go and research Fibonacci spirals in nature, shells, buildings, and paintings. You'll see those footprints everywhere.

Just as the Mandelbrot set reveals how beautiful complexity can emerge from simple formulas, our exploration of technology reveals the web of connections that bind code, circuits, behaviour, markets and psychology into one system.

The infinite potential of ideas equals innovation. We're navigating the boundaries of possibility in much the same way someone explores an infinite fractal zoom: you start at one point, zoom in, and the deeper you go, the more structure you discover. Each idea branches into multiple pathways, influencing your life and the lives of others.

Now imagine a world where everyone saw their ideas this way—as parts of an interconnected pattern, as nodes in a vast network of possibilities, as starting points rather than isolated events.

Interconnectedness fosters collaboration. Inventors, educators, traders, artists and dreamers can unite around patterns, not just opinions. Around numbers, not just narratives.

And that leads us straight to something that might make a lot of religious people—and a lot of scientists—equally uncomfortable:

The idea that mathematics might be the bridge between God and code. Between religion and technology.

## God Wrote the Universe in Mathematics

Galileo famously said that mathematics is the language in which 'God has written the universe.'

If you don't like the word 'God', swap it for Source, Nature, or Reality. The point stands.

Reality behaves mathematically.

Gravity follows mathematical laws. Planetary orbits trace mathematical curves. Light bends, reflects and refracts along equations. Electrons occupy orbitals according to formulas. Sound and vibration follow wave equations. Markets move according to probability, ratios and human behavioural patterns that can be modelled statistically.

Einstein, Tesla, Galileo, Leonardo da Vinci, Fibonacci, Newton—these weren't just 'smart people'. They all used formulas, ratios and geometry to describe phenomena that earlier cultures only described in myth, story and scripture.

Here's the interesting overlap:

Religious texts often hint at the same thing. They talk about an ordered universe. A 'word' (logo or symbol) behind everything. Number, weight and measure in creation. Mystical traditions like Kabbalah blend letters and numbers. In some interpretations, 'In the beginning was the Word' is not just poetic—it's an encoded statement about information, vibration and structure.

Now put that next to modern tech.

We build worlds in software using 0 and 1. We simulate physics with equations. We render 3D environments, light, shadow and motion with maths. We use algorithms—structured rule sets—to make decisions in finance, medicine, warfare and entertainment.

Religion says, "In the beginning was the Word."

Technology says, "In the beginning was the code."

Are these two completely different ideas? Or are they circling around the same principle from opposite sides?

## AI, Quantum Computing, and the Coming Convergence

Right now, as you read this, we're living in a time where artificial intelligence is writing code, analysing markets, generating images and videos, and even drafting books. Machine learning models detect patterns humans can't see unaided. Quantum computing is emerging with the potential to perform certain types of calculations at speeds that make today's supercomputers look slow.

We're scratching the surface.

We can already bring 100-year-old photos to life with animation. We can generate deepfake videos that blur the line between real and unreal. We have drones, tractors and aircraft navigating via onboard AI. We simulate pandemics, climate patterns and financial crashes.

If quantum computers mature, they could crack classical encryption by brute force in seconds and optimise complex problems in logistics, materials, and more in ways that look 'magical' compared to today.

This isn't centuries away. We're talking years, not decades.

Now project that forward twenty or thirty years.

Imagine a world where products are built using fractal principles, mimicking natural efficiency. Buildings self-regulate like ecosystems, following mathematical rules borrowed from nature. Sound and vibration are used deliberately in architecture, healing and personal performance. AI becomes not just a tool, but a collaborator. Quantum systems model reality in ways that eerily resemble some descriptions of 'higher realms'.

These worlds collide.

Technology, mathematics and what people call 'spirit' start to look less like separate departments and more like different lenses aimed at the same underlying pattern.

It's hard enough to fit that into one chapter. It's hard to fit into one book.

But if even a fraction of what we've talked about so far is accurate—even slightly—the way we look at life, history, and ourselves might not just be incomplete.

It might be backwards.

## Trading, Ratios, and Mental Constraints

Let's ground this back in something very practical: trading.

As a professional trader, I teach new traders to understand the maths of risk-to-reward instead of chasing some mythical holy-grail strategy that 'never loses'.

Most people believe you need a 90% win rate and a bulletproof entry system.

In reality, you can take a simple rule: risk 1% of your account per trade, aim to make 2% when you're right, and take the loss quickly when you're wrong. If you do that consistently, you only need to be right about 50% of the time to grow your account over hundreds of trades.

The maths is simple.

Win: +2.

Lose: –1.

If you take 100 trades and win 50, lose 50, that's (50 × 2) – (50 × 1) = +50 units. Adjust the strategy slightly: keep the 2:1 reward-to-risk ratio but improve your win rate to

60%. Out of 100 trades, 60 wins give +120, 40 losses give –40. Net: +80 units.

You're a profitable trader. No magic. Just math.

However...

Since diving deeper into human psychology—the same rabbit holes that led me to write this book—I've realised something else:

The math is necessary, but not sufficient.

There's more going on than ratios and formulas. There are beliefs, emotional set-points, and internal 'ceilings' on what we allow ourselves to earn, keep or risk.

I've seen traders with mathematically sound systems still blow up their accounts because some part of their psyche can't handle being right too often, can't let profits run, or can't take small, controlled losses without ego. It's like having a perfectly coded system running on a broken operating system.

The same pattern shows up between religion and science.

We can have perfect equations to describe gravity, but if we believe science and spirituality must be enemies, we'll never see that those equations might be describing the very order religious texts hint at.

We can have religious traditions talking about order, measure, 'divine law', symbols and wisdom, but if we believe maths is cold and godless, we'll never see formulas as another way of expressing that order.

What if the connection between math and God is more than poetic? What if the entire universe is governed by mathematics—and that's exactly what some of the more mystical parts of religion have been suggesting all along, just in a different language?

## Proving God with Math?

You'll find videos and arguments online trying to 'prove God exists using math.' Some draw on probability and fine-tuning, pointing out how unlikely a universe like ours would be if everything were random. Others argue that mathematics itself exists in some abstract, timeless realm—something we discover rather than invent. Some theologians and philosophers then say this timeless, ordered realm is one way of talking about the mind or nature of God.

Whether you agree or not is almost beside the point.

The point is that religion and science are, in some cases, starting to circle around the same questions using surprisingly similar tools: logic, probability, structure, and information.

Even if the jargon is different.

So, Where Do I Stand?

For me, I'm still very much on the fence when it comes to formal religion.

I'm not signed up to any single doctrine.

For me, this is less about blind belief and more about understanding—or at least trying to.

Maybe it's not that religion is wrong and science is right, or vice versa. Maybe we've just been stitching the story together in the wrong order.

Maybe technology is a modern ritual with silicon instead of stone. Maybe code is a new form of 'word'. Maybe mathematics is the bridge language between physical law and spiritual law. Maybe what ancient texts and builders were gesturing at with symbols, stories and stone, we are now rewriting in formulas and algorithms.

The important question for you is not, "Is this the final truth?"

The important question is: "If even a fraction of this is true, how should I live differently?"

Should you treat your thoughts as random or as lines of code?

Should you dismiss your ideas as 'just imagination,' or see them as potential blueprints waiting to be expressed in numbers, actions and structures?

Should you see markets as chaotic noise or as complex but patterned systems, you can learn to interact with both logically and emotionally.

Should you keep faith and logic in separate boxes, or allow the possibility that both are describing different angles of the same reality?

Technology, religion and numbers are not three separate worlds.

There's more to it, regardless of what order you put them in. The label matters less than your willingness to work with it instead of against it.

Because, once you see that the device in your hand runs on 0s and 1s; your brain runs on patterns and probabilities, markets run on fear and greed expressed in price, and the universe itself appears to run on equations... then the gap between praying, programming and planning your trades starts to look a lot smaller than you were taught to believe.

I recently was shown an interesting video called 'proving god exists using math'

https://www.youtube.com/watch?v=z0hxb5UVaNE

I'll leave it to you to go and watch.  Very interesting.

More crossovers

It's easy to forget that a lot of what we now call 'mindset work' was being decoded long before seminars, masterminds and Instagram quotes.

More than eighty years ago, Napoleon Hill spent twenty years interviewing over five hundred of the wealthiest people in America. Out of that work, he extracted what is, in effect, a mental algorithm. He believed it was hidden in plain sight in scripture—especially in the way the Bible talks about faith, prayer and "I AM".

Andrew Carnegie used it to build an empire. Henry Ford used a version of it to reshape industry. Earl Nightingale later condensed it into a single line: "You become what you think about." None of them credited luck. None of them blamed or praised circumstances. Every one of them operated from a particular state of mind that came before their results.

The Bible phrases it simply: "As a man thinketh in his heart, so is he." That isn't just poetic. It's mechanical. Your subconscious does not distinguish cleanly between a vividly imagined experience and a physical one. When you rehearse poverty and lack over and over, your nervous system learns to expect it. Your behaviour quietly follows that expectation. When you consistently think and feel in terms of growth and competence, your nervous system learns that instead.

Most of us were never taught this as a neutral law. We were given it as morality. We picked up the idea that humility means staying small. That 'spiritual' people shouldn't want material wealth. That money and God live in separate boxes. That cluster of beliefs has kept a lot of good people financially stuck, even as they work hard and try to do the right thing.

Hill's reading of scripture is very different. He points out that the same text that warns against greed also commands you to be fruitful and multiply. It condemns burying your talents. It says, in Mark 11:24, "what things soever ye desire, when ye pray, believe that ye receive them, and ye shall have them." The tension isn't in the verses; it's in how we've been trained to approach them.

We tend to pray with doubt. We ask due to fear. We visualise with a quiet, sceptical voice playing in the background, often out of desperation. The words we speak say one thing; the "I AM" we carry says another.

That clash is what Hill would call a mixed signal. Mixed signals don't create much. The universe, the market, the people around you—however you want to frame it—do not respond to your slogans. They respond to your dominant state.

Hill tells a story about a young man who came to him exhausted. He was working multiple jobs, saving every penny, and attending every seminar. His strategy list was long; his results were thin. Hill listened, then asked one simple question: "When you close your eyes and imagine your bank account, what do you see?"

The answer came fast: 'Bills. Debt. Stress.'

Hill didn't sugar-coat his reply. "Then that is what you will continue to experience. You are praying for abundance while rehearsing poverty."

He gave the man a specific assignment, based on what he had learned from Andrew Carnegie. Every morning, before he opened his eyes, the man had to see and feel a specific bank balance as already there. Not 'one day,' not 'I hope', but done. He had to repeat a precise statement of belief, with emotion, dozens of times. Then he had to move through his day as the man who already had that balance—not in delusional spending, but in posture, decisions and attitude.

In Hill's language, this wasn't playing pretend. It was activating what he called the law of belief.

Carnegie had given Hill a sentence that unlocks a lot of what you're reading in this book: "Faith is a state of mind which may be induced by auto-suggestion." If you let that land, it changes everything. Faith isn't a rare mood that occasionally descends on you from above. It's a state you can deliberately create.

Mark 11:24 says the same thing in another style: "When ye pray, believe that ye receive them, and ye shall have them." The whole verse pivots on the tense. "Believe that you receive." Present, not someday. For Hill, that verse is a technical instruction. When you pray or visualise from 'already received,' you stop arguing with yourself at the surface and start speaking directly to the deeper layer that actually runs your behaviour.

You can call that deeper layer what you like. Thomas Edison called it Infinite Intelligence. Hill called it the Divine Mind. In this book, in this book it's referred to as the Monad. The Bible calls it the Kingdom within. Different vocabularies, same idea: there is a deeper organising intelligence that responds to images and states held with conviction.

In Hill's story, the young man stuck with the exercise. Around three weeks in, he received an unexpected job offer that multiplied his income. A few days later, a long-forgotten debt was repaid in full. Within two months, the number on his statement matched the figure he'd been rehearsing every morning.

You can file that under coincidence if you want. Hill didn't. He saw it as law in action.

He boiled that law down to a simple formula that runs through 'Think and Grow Rich': desire, belief, expectation. Most people manage desire—they can tell you what they want. Very few train belief or expectation. Desire says, "I want to be rich." Belief says, "This is genuinely possible for me." Expectation says, "This is where I'm heading; my actions are just catching up."

The Bible doesn't say "Ask and hope." It says, "Ask, believing." Belief here isn't about getting a warm feeling once. It's a discipline. You build it the same way you build muscle: repetition under resistance. That's where Hill's idea of auto-suggestion meets what most people call prayer.

Auto-suggestion, in his terms, is you deliberately feeding your subconscious specific thoughts until they harden into automatic beliefs. Prayer, in the way he writes about it, is you deliberately turning your inner attention toward a higher intelligence and aligning your will with it. Different angles, same process.

When you pray, believing—really believing—you are using auto-suggestion in a spiritual frame. When you repeat "I AM" statements with reverence and emotion, you are praying. The energy is the same; the direction changes.

Hill also talks about transmutation: energy can't be destroyed, only transformed. Fear and faith are the same basic energy at different settings. His technique is straightforward. When fear enters, you don't pretend it isn't there. You acknowledge it, and then you deliberately introduce a stronger, conflicting thought of faith. The Bible's 'Perfect love casts out fear' becomes, in his language, 'A dominant thought of wealth displaces the dominant thought of poverty. You cannot hold two dominant thoughts at the same time. Choose.'

He read the biblical line 'Where two or three are gathered in my name, there am I in the midst of them' the same way. Not only as spiritual comfort, but as a description of a law of consciousness. Two or more minds genuinely aligned on a definite purpose create a third, intangible force greater than either one alone. That's the 'mastermind' principle.

This is why Carnegie built councils. Why Ford worked closely with Edison. Why serious entrepreneurs surround themselves with experts instead of just collecting followers. At the psychological level, they were deliberately building a field of shared belief.

Hill's advice is simple: choose a small group of people who actually share your vision; meet regularly; hold each other accountable not just for tasks, but for state. A mastermind group leaking doubt will quietly sink its members. A group that holds genuine, collective faith will multiply creative power.

Earl Nightingale took Hill's work and boiled it down further into 'The Strangest Secret': you become what you think about. He challenged people to thirty days of focused thought on a single, meaningful goal, held with conviction and backed by action. The surprising results he saw weren't magic. They were the same mechanism: clear desire, trained belief, daily expectation and movement.

By the time you reach this part of the book, you've met these themes from many angles: "I AM" as a command to the subconscious, faith as an internally generated state, visualisation as living in the end, mind and body as one system, Hill's language—desire, auto-suggestion, faith, mastermind—is just another dialect of the same thing.

You don't have to copy his rituals word for word. The point is to recognise that you are already doing versions of all this:

You are already using auto-suggestion every time you repeat "I am unlucky," "I am terrible with money," "I am just this way."

You are already praying every time you run a mental movie of your future and feel it as if it's true, for better or worse.

You are already building a mastermind every time you decide who you spend time with and what states they normalise.

The only real decision is whether you continue to do those things on autopilot—or whether you start doing them deliberately, in the direction of the life you actually want to live.

# Chapter 8: Motivation and Stoicism

**"We suffer more often in imagination than in reality."**

**Seneca**

Let's take it back a few steps.

In the first couple of chapters, we focused on mental strength: the wealth mindset, faith, belief, visualisation, and the science of attraction. Now we're going to weave in another strand that has been around for over two thousand years: Motivation and Stoicism.

Why bring them in?

Because all the mindset work in the world is useless if the first bad day throws you straight back into your old reality. If you can't hold your frequency when life punches you in the face, you don't really have a wealth mindset—you have a good mood. There's a big difference.

## Where Limits Are Really Built

Belief and faith—or more precisely, the lack of them—are where limitations are built.

When you genuinely believe in an end result, when you allow yourself to feel the emotions of already having it, and when you break that dream down into rational, actionable steps, something powerful dawns on you: most of what you took as 'impossible' was only ever a mental construct.

Not in the childish sense of 'I'll flap my arms and fly.' But in the adult sense that the walls you keep bumping into most of the time were designed in your head, reinforced by your upbringing, education, social circle, and the stories you repeat to yourself.

I don't believe the pyramids were just burial chambers. I do believe there is hidden knowledge—whether buried in archives, sitting in private collections, or coded in symbols—that never makes it into school books. I also believe the modern education system, for the most part, is designed to keep the mindset small and predictable.

Go to school. Learn to conform. Get a job. Pay your taxes. Fit into society. Retire. Die. Let your kids pay more tax on anything you leave behind.

Neat. Orderly. Contained.

What happens if you start nudging that script? Not with wishful thinking, but with clear, deliberate steps?

That's roughly where I found myself when this book began to form.

I'd already written and published two trading books with major houses. On the surface, they were about markets and strategies. The deeper purpose, though, was the same as this one: to help people see more of what's actually possible.

This book grew from the same intention—but the journey into it was different. I'd already spent years wandering through my own research: trading, psychology, ancient Egypt, metaphysics, science, religion, mathematics, and quantum physics. I'd spent thousands on books and courses and still found myself often in the same place:

"I don't fully get it."

That used to frustrate me. Over time, it became a strange kind of relief. Because the more I read, the more I realised that complete understanding isn't the point. What matters is knowing that you don't know—and staying honest about that.

Once I accepted that, another thought followed quickly: if I've gone deep into all this and still have more questions than answers, what chance does the average person—busy trying to survive—have of ever seeing these connections?

Why would they have spent thousands on books, psychology texts, Egyptology, quantum mechanics and religious philosophy? They haven't. They're making dinner, commuting or paying their bills.

So I asked myself a very simple question: what if I could pull together the parts that really matter—at least from my perspective—and condense them into one place?

That was the motivation for this book.

And that brings us to this chapter. Motivation and Stoicism are not minor add-ons. They're structural beams in the whole thing.

## When a Good Mood Isn't Enough

Creating a wealth mindset is one part of the puzzle. Using visualisation you actually believe in is another piece of the same puzzle. This builds emotional states you can project into the world. Not just cutting out pictures and glueing them to a board.

Without something deeper at the core—without a Stoic backbone and a real engine of motivation—setbacks will keep knocking you off your path. Sooner or later, they'll drag you right back to your default settings.

Here's a good example for you: think of an elite-level boxer.

You never hear a world champion stand in front of the cameras, claim to be 'the best in the world', and in the next breath admit they couldn't be bothered to train this morning. You don't hear them say, "My opponent's taller," or "He hits harder than me, so I'll probably lose."

Instead, you hear versions of the same thing: I've prepared. I've done the work. I'm ready.

They're not delusional; they're disciplined. Even if they've never read a page of philosophy, they are, in practice, Stoic. They focus on what they can control: training, diet, recovery, game plan, and mindset. They don't waste energy trying to shrink the other man's height or slow his punch. They accept what's given and then double down on their response to it.

When I started tying all these threads together for this book, something became clear: most people approach motivation the same shallow way they approach the wealth mindset.

They treat it like a New Year's resolution.

Go to the gym for a few weeks in January, and you think you've 'done health' for the year. Spend one evening feeling fired up about success, and you think you've 'worked on your mindset.'

Real life doesn't bend for that. Going to the gym for a month doesn't keep you fit in December. Thinking about success for five minutes does nothing if the next five days are spent complaining, procrastinating, and replaying old patterns.

Just like everything else in this book, the shift is more fundamental. It's about small changes, done daily, and taken seriously enough that when you slip—because you will—you realign quickly instead of spiralling for months.

Over time, that kind of approach does something you may not notice day-to-day. Motivation stops being a spike and becomes a baseline. Confidence stops being a mood and becomes a trait. Your posture, tone and presence begin to change long before the external results catch up.

You become the kind of person who is actually ready to receive what you've been asking the universe for.

That's why Stoicism matters here. It's not decor. It's load-bearing.

## The Old Philosophy for a Messy World

Stoicism gets thrown around a lot on social media these days—quotes on a background, a quick 'be unbothered' slogan—but the real thing has deep roots.

It began in Athens, early 3rd century BC, with a man called Zeno of Citium. He taught in a painted colonnade, the Stoa Poikile, which is how we got the name 'Stoicism'.

Over time, the philosophy travelled and evolved, hitting its stride during the Roman Empire. It became the private operating system for serious people doing serious jobs: politicians, generals, soldiers, emperors. Men whose days were full of war, disease, betrayal, economic chaos and backstabbing—not theoretical debates over coffee.

Seneca advised Nero and tried to live well in a court full of paranoia and violence. Epictetus began life as a slave, crippled, and still insisted that although your body can be chained, your mind is yours if you claim it. Marcus Aurelius, emperor of Rome, sat in military camps at night and wrote his journal—later known as 'Meditations'—to remind himself not to become arrogant, bitter, or weak while bearing the weight of an empire.

Stoicism is not about being a robot. It's not about killing feelings. It's about thinking clearly, acting with integrity, exercising self-control, and refusing to be thrown around emotionally by everything you can't change.

The Stoics believed real happiness—what they called 'eudaimonia'—comes from three things: seeing the world as it truly is rather than how you wish it was; practising virtues like courage, wisdom, justice and moderation; and putting most of your attention on what's actually in your control.

That last part changes everything when you really let it in.

## What You Can and Can't Control

The dichotomy of control is the beating heart of Stoicism.

Some things are within your control. Some are not. That's it.

You have control over your thoughts, your intentions, your choices, your effort, and your response. You do not have control over the weather, interest rates, your past, a stranger's opinion of you, random events, the traffic this morning, or the price of oil.

Most people live the other way around. They exhaust themselves emotionally over what they can't touch and neglect the one area they can rule absolutely: their inner state and their actions.

Stoicism flips that.

Stop trying to control the uncontrollable. Start mastering what is yours.

When you actually begin to filter your day through that lens, you waste less energy, you ride fewer emotional roller-coasters, and you have more bandwidth to build something meaningful.

## Where Motivation Usually Dies

Imagine you've just read the earlier chapters of this book. You're fired up. You go to bed the night before imagining your new future: money sorted, work that excites you, pressure eased. You wake up early the next day, determined to 'start living differently.'

You get dressed. You walk into the kitchen.

The toast burns. The coffee goes cold while you check your phone. You're already running slightly behind.

In that little gap, something crucial happens. Most people don't ignore it and move on. They narrate it.

"Typical. My day is already going wrong. Nothing goes right for me."

The words seem throwaway, but they're not. You've just told your subconscious to expect chaos. You've reaffirmed an identity: "things go wrong for me."

You get in the car. The petrol light is on. Now you're definitely cutting it fine. You swear under your breath. Your shoulders rise. Your breathing shortens. Every slow driver becomes an insult to your existence.

By the time you reach work, that powerful morning state you had—the visualisation, the sense of possibility, the wealth mindset—has been drowned in a puddle of burnt bread and low fuel.

From a Stoic point of view, nothing that happened is catastrophic. None of it is even interesting. The toast is already burnt; you can't unburn it. The coffee has already cooled; you can choose to heat it up or drink it or leave it. The petrol light is a fact that needs a simple action: refuel.

The only meaningful variable in that whole sequence is your response.

Do you let these trivial events carry your state away like leaves in a stream? Or do you treat them as noise, adjust behaviour where possible, and move on without dragging a story behind you?

From the 'frequency' angle we've been using in this book, what you did in that morning narrative was retune yourself. You started the day on one signal—gratitude, focus, possibility—and allowed a few small glitches to pull you back into scarcity, irritation and victimhood.

You did that with your own thoughts.

This is where Stoicism and motivation overlap perfectly. Motivation got you out of bed earlier. Stoicism would have allowed you to keep your signal intact.

## Risk Management for the Soul

If you trade, you're already familiar with risk management.

You know you need stop losses. You know you need sensible position sizes. You know that not every trade will win. You expect losing trades as part of the system instead of treating each one as a personal betrayal from the market.

Stoicism is the equivalent of risk management for your emotional life.

Instead of being shocked every time something goes wrong, you start to expect it as part of the terrain. People will be late. Technology will fail. Deals will fall through. Illness will appear. Loved ones will be difficult. You don't love this fact—but you accept it.

Seneca used a mental exercise called 'premeditatio malorum'—the premeditation of evils. He would occasionally imagine things going wrong on purpose: money lost, exile, failure, sickness. Not to dwell on negativity, but so that if any of those things did happen, he wouldn't be destroyed psychologically. He'd rehearsed his response.

This is not so far from what good traders and entrepreneurs do when they run worst-case scenarios. And it's not far from what I encouraged you to do earlier with visualisation: live in the successful end state, yes—but also see yourself handling setbacks with calm and clarity.

You can use your imagination not just to taste your eventual win, but to practise staying centred when the graph dips.

That's Stoic visualisation.

## What Motivation Actually Is

People mix up motivation with hype.

They think motivation is the adrenaline spike you get from a YouTube montage of private jets and supercars, or from someone shouting into a microphone on stage.

Motivation, in the way I've come to see it, is quieter. It's a combination of knowing why you're doing something, knowing roughly where you're heading, and being willing to move even when you don't feel like it.

Stoicism doesn't kill that. It protects it.

When those two work together, something elegant happens. Motivation gives you the reason to act. Stoicism gives you the spine to keep acting long after the initial excitement has faded.

Return to the boxer.

He is utterly driven to win. He's willing to run in the rain, to spar when tired, to say no to things most people would say yes to. But he also doesn't fall apart when he gets hit. He expects to get hit. He has trained his body and his mind to absorb the shot, adjust, and stay present.

Drive without discipline burns out quickly. Discipline without drive can survive hardship but go nowhere.

Together, they create a line: a direction, and the strength to keep walking in it.

That's exactly what you need when you set out to change your financial life, your health, your business, or anything else of significance.

Stoicism meets vibration.

Now, let's tie this back to everything else.

We've talked about the wealth mindset: removing doubt at the core level and living from the belief that your outcome is already set. We've explored how faith and fortune converge when you stop outsourcing your power and start acting as though your effort matters. We've covered visualisation as 'living in the end', and the science of attraction—how brain chemistry, body language and even quantum models can be used to understand why sustained focus changes outcomes.

We've zoomed out to ancient structures, frequencies in stone and sound, crystals as memory devices, and technology as modern ritual: circuits and code mirroring old ideas about symbols and law.

Now we're adding Stoicism: emotional stability, mental discipline, practical resilience.

Why does this matter for 'vibration' and 'attraction'?

Because your true signal isn't measured on the days when everything goes smoothly. The universe doesn't learn much about you when all the traffic lights are green.

Your signal shows itself when the plan fails. When the market snaps against you. When the client backs out. When the car breaks down, the same week the big bill lands. When someone you counted on lets you down.

In those moments, you have a choice: collapse back into the version of you that always lived on the edge of panic and resentment—or hold the line as the version of you that this book is trying to help you build.

Stoicism trains that second option.

You still feel the sting. You still feel the disappointment. But you don't become those feelings. You let them move through, and then you act from your principles, not from your first impulse.

Seen through the ‘frequency’ frame, the difference is stark. A person who only radiates confidence on good days sends a weak, inconsistent signal. A person who can maintain a steady, determined state through turbulence is broadcasting something entirely different.

That second person is the one who tries a new angle when the first doesn’t work. Who learns instead of sulks. Who adapts instead of surrenders. Who stays in motion instead of freezing.

The universe doesn’t reward wishful thinking. It responds to sustained alignment.

Motivation starts with that alignment. Stoicism holds it.

## Bringing It Into Your Life

So how do you actually bring Stoicism and motivation into your day in a way that supports everything else in this book?

You begin by noticing, in real time, what you reach for when something small goes wrong. Instead of automatically letting your mind run the old script—“Typical, my luck, this always happens to me”—you treat those moments as training drills. You catch yourself, take a breath, and choose a different response.

You start expecting, in advance, that the path you’re taking will involve friction. Before you launch a project, start a new business, or commit to a health plan, you take a few minutes to ask yourself: “What could realistically go wrong along the way? And when it does, who do I want to be in that moment?" You're not inviting failure; you're refusing to be surprised by it.

You stop waiting to 'feel motivated' and start moving because you've decided to. If you only work on your goals when the mood is perfect, you're not motivated—you're entertained. Motivation becomes a decision, not a sensation.

And gradually, you combine this with the visualisation work from earlier chapters. You don't only picture yourself at the finish line; you see yourself staying steady at mile three when it starts to rain. You imagine your future self handling a losing trade with calm, responding to criticism without shrinking, sticking to the plan when quick wins don't appear.

The more you rehearse that in your mind and then act it out in small ways—burnt toast, low petrol, awkward email—the stronger that inner muscle gets.

## Solid in a Shaky World

Stoicism is not about cutting yourself off from life. It's about staying solid in a world that is constantly shaking you.

Layer that onto a genuine wealth mindset, real faith in yourself, powerful visualisation, and an understanding of how your inner world shapes your outer results, and something important shifts.

You stop being a passenger, thrown from side to side by every bump in the road.

You become the driver.

And that is when the universe, the market, the people around you—however you want to frame it—finally has something consistent to respond to.

# Chapter 9: Mind and Body Connection

**"Whatever you dwell upon, you will draw to you."**

Bob Proctor

The intricate interplay of mind and body is often talked about but rarely understood.

In our pursuit of success, there is an undeniable connection between how we think and how we feel physically that deserves far more attention than it usually gets. Just as our thoughts shape our beliefs and perceptions, our physical state quietly shapes how we think and feel about ourselves and the world around us.

This interplay is not just philosophical. It has very practical implications for achieving goals, manifesting desires, maintaining a wealth mindset, and holding your 'frequency' when life pushes back.

For me, this is one of the chapters where everything really starts to come together. You stop seeing 'mindset', 'health', 'emotion', 'motivation', 'success' as separate folders and start seeing one integrated system.

As you begin to understand the mind–body connection, you start to notice how physical health affects mental clarity and emotional resilience; how mental patterns and beliefs literally alter your biology; how the way you move, breathe, eat, and sleep all feed back into your mindset and 'signal'.

If you want success by design—not by accident—you have to recognise that your thoughts and feelings are tied to your physical existence. Neville Goddard talked about this repeatedly: "feeling is the secret." Feeling is not some vague, mystical idea. Feeling is how your mind talks to your body.

## What Is the Mind–Body Connection Really?

The mind–body connection is simply the relationship between your mental processes and your physical health.

At its core, it says this: your thoughts, emotions and mental states can significantly influence your physical wellbeing, and your physical state can, in turn, affect how you think and feel.

That's not just an opinion. Psychology, neuroscience and medicine all converge on the same point: you are not a brain carrying a body around like luggage. You are one system.

This is where things became clearer for me. Once you stop treating 'brain', 'hormones', 'nervous system', 'immune system' as separate silos, you realise they are in constant conversation.

Your brain can be rewired—that's neuroplasticity. Hormones and neurotransmitters act as chemical messengers influencing mood, energy and focus. Your nervous system swings between fight-or-flight and rest-and-digest states. Your immune system responds differently depending on your stress levels and emotional climate.

All of them are talking to each other, all day long.

## Neuroplasticity: The Brain That Listens to Your Thoughts

Neuroplasticity is one of the clearest examples of the mind–body link. It's the brain's ability to reorganise itself by forming new neural connections throughout life.

In plain language, your brain is not fixed. Your 'wiring' is not destiny. What you repeat, you reinforce.

Thoughts are not just ghostly things that float in your head and disappear. Repeated thoughts strengthen particular circuits. Repeated feelings strengthen particular networks.

If you practise visualisation, affirmations and particular emotional states long enough, your brain begins to wire around them. Confidence becomes easier. Staying calm under pressure stops being a heroic effort and becomes more natural. Optimism stops feeling fake and starts to become your default lens.

The opposite is also true. Rehearsing failure, fear and doubt rewires your brain too. Constantly replaying worst-case scenarios strengthens anxiety pathways. Putting yourself down on repeat trains your brain to filter for proof that you're not enough.

The brain does not have a moral opinion about this. It doesn't say, "This is a good belief, I'll reinforce it", and "This is a bad one, I'll block it." It simply responds: "What you repeat, I will automate."

This is part of why Neville's idea of 'living in the end' is so powerful. When you repeatedly feel and rehearse the state of already being the person you want to be—wealthy, healthy, confident, calm—you're not just 'pretending.' You are sending clear instructions to your nervous system: wire this in; make this normal.

The fascinating part is that your brain doesn't clearly distinguish between vivid imagination and external reality the way you might assume. fMRI scans show that imagining an action lights up many of the same regions as actually performing it. Athletes who mentally rehearse their performance often see measurable improvements. Even 'imagined practice' can grow skill, because the brain is training timing and pattern.

Later in this book, when we look at hypnosis and suggestion, this will become even more obvious. You'll see that your brain has been accepting suggestions—from parents, school, media, culture, and from you—since day one. The only question is whether you're now willing to become conscious of what you're installing.

## Emotions in the Body: Your First Feedback System

If you want quick proof that the mind and body are one system, you only need to look at how emotions show up physically.

Think about how stress feels. Tight shoulders, shallow breathing, jaw clenched, knots in your stomach. Anxiety? Fast heart rate, restless legs, sweaty palms, a sense of agitation in your chest. Happiness? Warmth in the chest, softer muscles, an open posture. Shame? Your chest sinks, your gaze drops, your body feels heavier.

Emotions don't just sit in your head. They appear in your flesh.

Once you recognise that, you gain a new level of control. You can reverse-engineer your state. You can notice that your shoulders are up near your ears and ask, "What am I telling myself right now?" You can consciously relax your jaw, drop your shoulders, slow your breathing, and feed calm back into your system.

This is where Stoicism meets biology. You don't always control the trigger—that email arrives, that person says what they say, the bill lands when it lands. But you do control how you respond. Changing your body changes what your mind has access to.

This is why I keep saying this is not "think and it will happen." It's: use thought and body deliberately, in sync, over time—and watch what shifts.

## The Placebo Effect: Belief That Changes Flesh

The placebo effect is one of the most astonishing demonstrations of the mind–body connection.

A placebo is simple: a person believes they're getting a real treatment, whereas in reality, they receive something inert—a sugar pill, a saline injection and yet they still improve. Not just in their imagination. In many studies, measurable physiological changes occur. Pain markers drop. Symptoms reduce. The brain and body respond as if a real intervention is happening.

What changed?

Belief. Expectation.

The story in the mind.

The mind sends a strong enough 'this is helping' signal, and the body responds in line with that instruction.

Pause on that for a moment.

If belief in a fake pill can alter pain levels, immune responses and healing speed... what might sustained belief in your own future, your own ability, your own worth do over months and years?

The darker side of this is the nocebo effect—the flip side of placebo, where negative expectation makes things worse. "This will hurt. This won't work. I always get ill." And the body obliges.

Think about a child's imagination. They fly with dragons, build whole worlds from Lego, speak to invisible friends, become astronauts, warriors, magicians. Their nervous system is wide open to possibilities.

It's only when adults come along with "be realistic" and "that's not how the world works" that the shutters start closing. That imaginative muscle—so potent for shaping belief and, by extension, biology—starts to atrophy.

The mind is an incredibly powerful tool. But without a reasonably well-maintained physical base, it cannot operate properly. Chronic stress reshapes brain structures. Inflammation affects mood and clarity. Exhaustion knocks out your ability to hold any kind of elevated state.

Mind and body are not two separate things. They are one loop running in real time.

## Your Body as Hardware, Your Mind as Operating System

A useful way to picture it is this: your body is the hardware; your mind—both conscious and subconscious—is the operating system.

If the hardware is overheating, clogged, underpowered, and permanently stuck at 2% battery, you can install the best OS in the world, and it will still freeze and glitch. On the flip side, if the operating system is corrupted by limiting beliefs, fear loops, shame and self-hatred, you can have excellent hardware and still dramatically underperform.

To run at a high level, you want both: a healthy body to support energy, focus and resilience, and a well-trained mind to direct that energy intelligently.

So let's make this tangible.

## Food: Programming Through Fuel

Everything you eat is doing one of two things: it's either helping your system run clean and clear, or it's adding friction, fog and volatility.

Nutrient-rich food supports brain health and cognitive function. Poor nutrition does the opposite. Healthy fats help build and maintain brain structure and signalling. Adequate protein provides the raw materials for neurotransmitters like dopamine and serotonin. Vitamins and minerals support energy production, immune function and focus. Highly processed food, constant sugar hits and endless snacking push your blood sugar up and down like a rollercoaster, dragging your mood and concentration along with it.

If your diet is chaotic, energy swings start to feel 'just the way you are.' Mood swings feel like your personality. Brain fog feels like your identity. You start calling yourself lazy, unmotivated, not very sharp—when in reality, your fuel is working against you.

You wouldn't pour the cheapest, dirtiest fuel into a high-performance car and then blame the engine when it splutters. Yet people do the equivalent to their bodies every day, then use the resulting chaos as evidence of their own inadequacy.

I'm not here to give you a strict meal plan. I'm here to point out that if you want your mind to function at a high level, you need to start eating like someone who respects their brain. Hydrate as it matters—because it does. Notice how you feel after certain foods, and be honest about the patterns.

You're not just feeding a body. You're feeding your capability.

### Movement: Exercise as a Mental Upgrade

Regular movement is one of the most underused performance enhancers that exists.

Exercise releases endorphins, your body's natural mood lifters. It increases BDNF (Brain-Derived Neurotrophic Factor), which supports the growth of new brain cells and connections. It improves blood flow and oxygen delivery to the brain, boosting clarity, reaction time, and focus. It helps regulate sleep, appetite and stress hormones.

Look at it through the 'vibration' lens: when you sit hunched at a desk all day, your energy contracts. When you walk, lift, stretch, and breathe deeply, your energy expands. You shake up stagnant patterns—physically and mentally.

You don't need to live in the gym or chase a six-pack. But some form of regular movement that raises your heart rate, loads your muscles and challenges your lungs is non-negotiable if you're serious about performing at a high level in any area: business, leadership, creativity.

We're not talking about beach photos. We're talking about building a body that can carry a big vision without folding under it.

## Sleep: The Night Shift That Rewires You

Sleep is where your body and brain clock in for the night shift.

While you sleep, your body repairs tissues, muscles and organs. Your brain consolidates memories, processes the day and integrates learning. Waste products that accumulate in the brain during wakefulness—including some linked to cognitive decline—are cleared more efficiently.

Cut that process short and the bill arrives quickly. Cognitive performance drops. Emotional regulation falls apart. You're more irritable, more reactive, more prone to cravings and poor decisions. Your immune system takes a hit.

You can visualise wealth, repeat affirmations and buy every course you can find. But if you're living on four or five hours of sleep, hammering caffeine to stay upright, and scrolling a blue-lit screen until 1 a.m., your nervous system is in survival mode, not creation mode.

The simplest, fastest way to improve almost everything about your performance is to treat sleep as sacred. Often, adding just an extra hour or ninety minutes of quality sleep per night, consistently, changes mood, focus and resilience dramatically.

You wouldn't run your phone at 5% battery all the time and expect it to last the day. Don't do that to yourself and then blame your 'discipline' or 'mindset' when you feel wrecked.

## Meditation, Breath and Being in Your Body

Now we move into practices that directly bridge mind and body.

At its core, meditation is attention training. Most people's attention is dragged around all day by notifications, worries, and random thought loops. They are passengers in their own head.

Meditation asks you to sit down, pick something simple to focus on—your breathing, a sound, bodily sensations—and notice, without judgement, when your mind wanders. Then you gently bring it back. That's the whole game.

Over time, this simple practice changes you. You become less reactive. You notice thoughts and feelings before they run you. Brain scans show structural changes in areas linked to attention and emotional regulation. You move from being pushed around by your mind to steering it.

Breathwork is another direct lever.

Slow, deep breathing—especially into the diaphragm—activates your parasympathetic nervous system, the 'rest and digest' state. Your heart rate slows. Your muscles release. Your body receives the message: we are safe.

You cannot instantly stop an external event. But you can, at any moment, decide to take ten slow breaths. As your physiology calms, your mental options expand. Your breath is a remote control for your nervous system that most people never bother to pick up.

Practices like yoga, tai chi and qigong combine movement, breath and focused attention. They train you to be present in your body, to notice how you stand and move, where you hold tension, and how your internal state changes under stress. They also teach you to stay calm in discomfort—an incredibly useful skill in trading, business, or any high-pressure situation.

Seen through the 'signal' metaphor, these practices are clear static. They ground scattered mental energy. They create a more stable, coherent internal field. You become someone who feels better, moves better, and thinks better—not by accident, but by practice.

## Emotional Intelligence: Reading and Steering Your System

Emotional intelligence is simply your ability to notice, understand and manage your own emotions, and to recognise and respond effectively to the emotions of others.

Why does that matter for wealth and success?

Because the world of money is not just numbers on a screen. It is humans in motion. Deals are made and lost based on trust, fear and greed. Teams thrive or fracture based on communication and emotional safety. Clients, partners and investors say yes or no largely based on how they feel around you.

If you have no awareness of your own emotional patterns, if you can't regulate yourself, if you miss the emotional cues other people are broadcasting, you're going to repeat the same sabotaging cycles, no matter how good your technical skills are.

Mind–body awareness feeds emotional intelligence directly. You notice a knot in your stomach and ask, "What story am I running?" You feel a sense of expansion and ask, "What about this feels aligned?" You start to read others' posture, tone and micro-expressions as part of the equation instead of just listening to their words.

## Building the Platform for a Wealthy, Aligned Life

The point of all this is straightforward: the better you understand and support your physical state, the more powerful your mindset work becomes.

If you eat in a way that keeps your energy smooth, if you sleep deeply, move regularly, hydrate properly, breathe consciously and give yourself some quiet mental space, you are building a platform. On top of that, you can then layer affirmations, visualisations and belief work.

Suddenly, your affirmations aren't fighting brain fog. Your visualisations aren't trying to land in a nervous system running on fumes. Your wealth mindset isn't being undermined by a body that feels like it's always on the brink.

This combination—solid physical foundations plus deliberate mental training—heightens self-belief. It makes the action more natural instead of forced. It makes it easier to stay in a higher, more coherent state. Mental rehearsal then has a healthy, responsive body to work through.

Your physical state and your aspirations stop arguing and start collaborating.

## The Process, Not the Magic Switch

If there's one thing I want you to take from this chapter, it's that none of this is about a magic switch. It's about a process.

You improve your sleep slightly. You wake a little clearer. You make slightly better choices. You handle stress a touch more calmly. That experience builds a bit of confidence. That confidence nudges you to move more, eat slightly better, tackle a task you've been putting off.

As your physical state improves, your confidence grows. You understand concepts more easily. New knowledge starts to stick. Your whole presence—what some people would call your aura—begins to shift.

People feel that, often, before you do. They respond differently. New opportunities appear, or old ones become visible in a way they weren't before. That response from the world reinforces your new identity, which in turn shifts your mindset to expect and allow more good things.

Everything is connected.

## Memory and Imagination: One Circuit, Two Directions

One of the most striking things I found when digging into the neuroscience was this: memory and imagination run on many of the same circuits.

When you vividly recall something from your past and when you vividly picture something in your future, your brain lights up overlapping regions. It fills in gaps, colours in details, and writes the story. To your nervous system, both are 'real enough' to influence your present.

Your brain cannot cleanly and consistently tell the difference.

That has enormous implications.

If you repeatedly imagine a life where you are healthy, strong, wealthy and calm, your brain begins to treat that as part of its internal map of what's possible and normal. If you repeatedly imagine embarrassment, failure, bankruptcy and rejection, it will wire that as your expected, familiar world.

When you picture your ideal life, honestly, what do you see? Are you slumped on a sofa, exhausted, sick and angry at the world? Or are you healthy, energised, engaged, and moving with purpose? Are you bitter, resentful and stuck—or smiling, grateful, confident and expansive?

You know the answer.

So why would you live physically in a way that contradicts the life you claim to be visualising? Why imagine yourself vibrant and powerful, then feed your body like you hate it, sleep like it's optional, and move only when forced?

When you start rewriting your pre-programmed mindset, and you support that rewrite with new physical behaviours, everything around you begins to shift. Your beliefs shape what you allow yourself to experience. Your body either supports those beliefs—or constantly drags you back toward your old identity.

## Don't Be a Piece on the Board

The final image I want to leave you with is this: don't live like a pawn on a chessboard, pushed around by invisible hands. Be the player.

That means stop treating your body as an inconvenient vehicle you drag behind you, and stop treating your mind as something that just 'happens' to you. Start seeing both as tools you can train, align and direct.

Reality, for the most part, is created—not chased. You create it with the thoughts you rehearse. You create it with the emotions you practise. You create it with the way you treat your own hardware. You create it with your response to the inevitable noise and chaos.

The mind–body connection is not just an inspiring phrase. It's literally how you're wired.

Once you understand that, you stop waiting for change to arrive from somewhere out there...

...and you start building the system inside you that makes change inevitable.

# Chapter 10: The Hidden Circuit

**"There are no secrets in life; just hidden truths that lie beneath the surface."**

**Michael C Hall**

By the time you reach this point in the book, you've probably noticed a pattern running quietly underneath everything:

Your thoughts affect your body.

Your body affects your thoughts.

Emotion is energy in motion.

Structure, rhythm and intention shape that energy.

We've explored that in visualisation, and in building a wealth mindset, and even in the way ancient builders might have used sound and geometry. In this chapter, I want to show you how the same logic appears in practices that, at first glance, can look soft, mystical, or even a bit ridiculous.

The first time I saw tai chi in real life, I mentally filed it under 'retirement-home yoga.'

It was in a park. A small group of older people in loose clothing moved slowly, hands drifting through the air like they were underwater. It looked peaceful enough, almost like a moving meditation. But powerful? It didn't exactly scream 'edge.' I walked past, the way you probably have, thinking, "Nice stretching... but what's that got to do with anything real?"

For years, that was the end of it for me.

Later, as I went deeper into the mind–body connection, into frequencies and fields, I started to see tai chi differently. Once you get past the clichés, it becomes one of the clearest physical demonstrations of something this entire book has been trying to say:

Intention directs energy.

Structure channels force.

The nervous system can be trained to work with, rather than against, the environment.

And that is exactly what you are trying to do in your trading, your business, and your life: stop fighting reality blindly and start moving with it.

## Seeing Tai Chi Through a Different Lens

Strip away the mystical language for a moment and imagine you're watching tai chi as an engineer.

A group of people is moving in slow motion. Their feet glide almost silently across the ground. Hands spiral, circle, extend, and withdraw in precise lines. Their attention is not out there on their phones or drifting into the distance; it's clearly anchored inside their own bodies. Every shift of weight, every turn of the waist, every breath feels like part of one unbroken chain.

Now imagine you filmed them and then played the footage back at three or four times the speed. Suddenly, those gentle arcs start to look suspiciously like a fight.

And originally, that's exactly what tai chi was.

Those soft, rounded movements are strikes, locks, throws and deflections—stretched out in slow motion so the practitioner can feel everything that happens between the thought 'move' and the moment of impact.

The aim is not 'stretch and relax.' The aim is: align your structure, root your weight into the ground, coordinate your entire body as one unit, and direct force from the floor, up through the legs, driven by the waist, expressed in the hands.

In simple terms: stop fighting your own body so you can deal intelligently with whatever comes at you from outside.

That logic should sound familiar. It's the same logic you need for managing risk, handling stress, or building a wealth mindset. Most people lose, not because the world is too strong, but because they are wasting half their strength fighting themselves internally at the same time.

The 'energy manipulation' in Tai Chi isn't wand-waving. It's a disciplined way of teaching your brain, fascia, muscles and nervous system to behave differently under load. When you feel it—or stand opposite someone who is genuinely skilled—it does feel as if you've stepped into strange territory.

## What 'Energy' Really Means Here

The word 'energy' gets abused. In tai chi and other Chinese internal arts, you'll hear about qi (or chi), often translated as 'vital energy' or 'life force.' Western ears often shut down at that point, because we're used to thinking of energy as calories on a label or numbers on an electricity bill.

Here's a more grounded way to look at qi in this context:

Qi is simply the way your body organises and moves breath, tension, relaxation, awareness and force through its structure.

When a tai chi teacher talks about 'moving qi,' they are usually doing three very practical things at once: they're guiding where you place your attention, how you distribute tension and relaxation, and how you initiate movement from your centre rather than flapping isolated limbs around.

If that still sounds abstract, picture two versions of yourself.

In the first, you're stressed. Your breathing is shallow and high in the chest. Your shoulders are creeping up towards your ears. Your weight feels like it's sitting in your throat and neck. Your thoughts are sharp, choppy, and relentless.

That's one qi pattern.

In the second, you slow your breathing on purpose. Your shoulders drop. You feel the soles of your feet on the floor. Your stance widens slightly. Your gaze softens. You move your arms slowly and you can feel how your weight passes through your legs as you turn.

Inside, something changes.

Same body. Same bones and muscles. But an entirely different internal distribution of 'energy'.

Tai chi is essentially the art of learning to choose that second pattern on demand—and then hold it when you're under pressure.

Again, sound familiar?

It's the same thing we spoke about with Stoicism and trading: keeping your signal stable even as markets spike, headlines scream, or social media panics.

## Rooting, Centre and Song

In practice, tai chi drills a few key ideas into you so deeply that they stop being concepts and become instincts. They are worth stealing.

The first is rooting.

A rooted person doesn't strike a heroic pose or look tense. They actually look relaxed. But when you place your hands on their chest or shoulders and press, something strange happens: instead of swaying or stiffening, they seem to 'send' your force straight down into the ground. Their upper body remains relatively calm while your own balance starts to wobble.

They feel like a tree: flexible at the top, solid at the base.

If you're not rooted, your upper body is doing all the work of carrying your weight. Your legs are passengers. A child could knock you sideways—physically, emotionally, or even financially.

Translated into daily life: Are you grounded or jittery? Do small events knock you flat? Does one setback send you spinning emotionally as if you had no base beneath you?

Rooting, in tai chi terms, is 'energy manipulation' in the sense that you repeatedly move, sink, and align your body until your nervous system treats 'grounded' as normal. The next time someone raises their voice at you, or a position moves hard against you, you'll still feel the jolt of adrenaline—but you won't topple as easily. Your feet stay under you. You respond instead of flailing.

The second principle is dantian: your centre.

In tai chi, the dantian is an area a few inches below the navel, deep inside. It's treated as the body's physical and energetic centre. Every proper movement starts there. Hands and feet are messengers. The waist and hips turn, the centre shifts, and the limbs arrive last.

In business, this is like making sure every action flows from your core principles and objectives, not from random impulses, fears, or external noise.

When your 'centre' is unclear, you chase everything. One day, you're committed to one strategy; by the end of the week, you've abandoned it because someone louder on social media said something different. You end up exhausted and nowhere.

When your movements—on the mat, on the charts, or in a conversation—come from a clear centre, people sense it. In tai chi, they literally feel it if they push on you: their hands meet a unified mass rather than disconnected parts. In life, it reads as confidence and presence.

Energetically, this isn't magic. It's the trained habit of constantly checking in with your middle: Where am I? What am I doing? Why now?

The third principle is song.

Song is often lazily translated as 'relaxation,' which is misleading. Say "relax" and most people go floppy, unfocused, half asleep. That's not it.

Song is relaxed but ready.

Your joints are open. Your muscles are not rigid, but they're not slack either. If something happened, you could act explosively without needing a big 'wind-up' to unlock your own tension first.

Think of a cat on a windowsill. It appears melted across the ledge, eyes half closed, completely casual. But if a bird lands in the garden, in one smooth blur, that same animal is up, coiled and mid-air. It didn't have to unfreeze first. It was soft, but primed.

In energy terms, song means you're not wasting strength bracing against imaginary threats. You're not gripping everything with white knuckles just in case. Because you're not jammed, you can feel subtle changes around you.

In markets, song looks like this: you're not clenching over every tick. You're not obsessing over a single trade as if your entire life hangs on it. You're relaxed enough to see the bigger structure developing, and sharp enough to act decisively when your setup actually appears.

The 'energy work' here is the continuous letting go of what you don't need, so what you do need can move through you without friction.

## Intent: The Invisible Line

There's another layer that connects all of this directly back to the rest of the book: intent.

Chinese internal arts distinguish between li (raw, mechanical strength), qi (integrated internal power), and yi (intent—the mind directing where the energy goes).

A beginner in tai chi operates mostly in li. They move their arms and legs. From the outside, what they're doing roughly matches the shape of the form, but it's disjointed and heavy.

With time, they discover qi. Movements originate in the legs, twist through the waist, and reach the hands in a connected wave. The form looks smoother, more efficient, more powerful—even at slow speeds.

At a higher level, the practitioner thinks less about the external shape and more about yi—the thin, invisible line of intention. They feel the direction of force, the path through space, the other person's balance. The body follows that mental line almost automatically.

This is not that different from what happens when you visualise a trade, a conversation, or a life outcome with clarity and repetition. The clearer and more consistent your intent, the easier it is for your actions to line up with it.

Energy manipulation at this level looks like focused attention (yi) travelling through a coherent internal structure (qi), expressed via a relaxed, responsive body (song), to create a very physical result: stability, redirection of someone else's force, efficient motion.

Translated into ordinary life: a clear goal, beliefs and habits that match that goal, and flexible but disciplined behaviour. That combination moves you from a random, reactive existence into a life that feels increasingly designed.

## Push-Hands and the Market

Before we move on to two more tools—mudras and Dim Mak—there's one more tai chi practice worth mentioning: push-hands.

Two people stand facing one another, palms or forearms gently touching. They begin to move together, circling, yielding, pressing. The aim is not to fight aggressively. It's to feel.

Where is their weight? Where are they stiff? Where are they hollow?

As you stay in contact, you apply pressure in tiny spirals, listening with your hands. If you sense a gap in their structure—a place they've overcommitted—you angle them gently into that emptiness. Often, with almost no apparent effort, they stumble, step, or suddenly lose balance.

It's a conversation in structure and force.

As you improve, you learn not to fight force head-on. You learn not to respond to hardness with more hardness. Instead, you yield just enough, redirect just enough, and let the other person's own tension undo them.

Zoom out, and you'll see people doing something similar to stiff, ego-driven founders every day.

Founders will say 'my company' is worth X, and the investor will say, "no you think your idea is worth X. The reality is it's not worth Y"

Your internal rigidity turns every move into an insult. You start revenge-talking, fighting the investor verbally, but they don't care that you even exist. Your own tension is used against you.

If you behave more like a skilled push-hands player, you accept what is. You yield where it makes sense. You apply pressure when you feel real weakness, not imagined insult. You stay connected to the market, but not emotionally attached to any one prediction.

You're still 'manipulating energy,' but not by trying to bend reality to your will. You're manipulating it by refusing to be baited into over-reactivity.

That is real power.

## Changing State Through Shape and Breath

All of this loops back into the mind–body connection we explored in the previous chapter.

Tai chi becomes a living laboratory where you can watch the cycle in action.

Think fearful, victim-style thoughts, and your body responds: breathing tightens, posture collapses, balance goes. You become easy to move—on the mat, in meetings, on the charts.

Think grounded, confident thoughts, and your body reorganises: lungs open, unnecessary muscle tension drops, your stance widens. You become much harder to knock off course.

With practice, tai chi lets you run that process in reverse as well. Change how you move, and your brain waves start to shift. Alter your breathing rhythm and your heart rate follows; your thoughts slow down; your perception widens.

You are literally changing your internal state by adjusting how you stand, walk and inhale.

Hypnosis, which we'll get into in the next chapter, does something similar using words and suggestion. Tai chi does it using posture and motion. Both are different interfaces for rewriting the same underlying software.

## Mudras: Power in Your Hands

Tai chi is one expression of this mind–body energy language. Another, much smaller but surprisingly potent one sits right in front of you: your hands.

You've probably seen mudras without realising what you're looking at. A politician on stage, fingers steepled. A CEO in an interview, hands forming a loose triangle. An athlete celebrating, palms pressed together. People in prayer. Repeated, specific shapes.

Some parts of the internet love to frame these as secret society codes or Illuminati signals. It makes for exciting videos. In reality, the truth is simpler and, to me, more useful.

Mudras are hand gestures that seem to direct attention and, quite possibly, neural and electrical flow in particular ways.

Consider how dense your hands are with nerve endings. Think about how often you 'talk with your hands' without planning it. Your hands are directly wired into your brain. When you press, join or shape your fingers in certain patterns, you change muscle activation, tension lines, where your mind naturally settles, and how your body feels.

Certain mudras have been used in yogic and meditative traditions for thousands of years. You don't have to swallow every mystical claim about chakras and prana to acknowledge that these postures do something.

Take Gyan mudra, sometimes called the gesture of knowledge: the tip of the index finger touching the tip of the thumb, other fingers gently extended. Hold that position while breathing slowly and focusing, and many people report a natural sense of clarity, of 'dropping into' thinking mode. You're giving your nervous system a familiar loop of stimulation that it associates with attention and centred awareness.

Or Anjali mudra—palms pressed together at the heart, thumbs resting against the chest. Spend a minute there and notice the feeling of symmetry and calm that often arises. You are literally connecting both sides of the body across the midline, feeding equal sensory input to both hemispheres of the brain. It feels like unity because, at a basic level, that's what you're creating.

Then there's Hakini mudra—fingertips of both hands touching lightly in front of you, as if you're holding a small invisible ball. People naturally adopt this when they're thinking deeply or planning. Some studies suggest it encourages communication between the brain's hemispheres. You might catch yourself doing it unconsciously when you're visualising a future or designing a trade.

The point is that small, deliberate changes in how you hold your hands can shift your state—towards focus, calm, or alertness. They are physical switches you can flip to trigger mental modes, the same way a tai chi stance triggers a feeling of being grounded.

Ancient teachers wrapped this knowledge in chakra diagrams and talk of prana. Modern scientists might talk instead about proprioception, neural loops and bilateral integration. The vocabulary differs; the principle does not.

Your body is full of such switches. When you learn to use them consciously, you 'manipulate energy' in a way that's anything but vague: your heart rate drops, your breathing smooths out, your voice steadies, your thinking sharpens.

### Dim Mak: The Darker Map

If tai chi and mudras are the softer sides of this knowledge, Dim Mak sits at the harsher end.

I was fortunate—or unfortunate, depending on your view—to be exposed to small pieces of Dim Mak years ago. The name itself sounds like something from a bad martial arts film: 'death touch,' secret strikes that kill with a tap. A lot of what's said about it online is nonsense.

Underneath the drama, though, lies a blunt reality.

The human body is a structured, patterned, energetic system. If you understand the structure, you can influence that system—for healing or for harm.

Ancient Chinese medicine divided the body into meridians: routes where qi was said to flow. Acupuncture and acupressure use that map to relieve pain, rebalance and heal. Tai chi uses it to circulate and harmonise.

Dim Mak uses the same map with a very different intention.

Instead of gently stimulating a point to ease a headache, you strike it at a specific angle, depth and timing to create maximum disruption: unconsciousness, organ shock, temporary paralysis.

Strip away the old qi descriptions and modern medicine quietly agrees on the underlying mechanics. There are points on the body where nerves, blood vessels and vital organs lie close to the surface. A sharp blow to the carotid sinus in the neck can cause blood pressure to drop suddenly and the person to black out. A strike to the solar plexus can spasm the diaphragm and knock the wind from your lungs. A well-timed impact over the heart can trigger a fatal arrhythmia—commotio cordis. A punch to the jaw can spin the head and shock the brainstem, dropping someone instantly.

Dim Mak is, in many ways, a pre-scientific map of these vulnerabilities. The old masters wrapped their observations in qi language and passed them down through forms integrated into arts like tai chi, bagua, xingyi. They emphasised three things above all: angle, depth and timing.

The 'energy manipulation' here is not some cinematic finger flick from across the room. It's the practitioner's ability to stay in song so that power travels cleanly through their structure; to stay rooted so the ground, not just the arm, powers the strike; and to focus yi—their intent—on a precise point at the precise moment.

From the outside, it can look almost lazy. From the inside, it is structure plus timing plus anatomy, resulting in a disproportionate effect.

Here's the interesting twist: the same point that a Dim Mak striker uses to drop someone may be the one an acupuncturist uses to relieve pain or restore balance. Same body. Same map. Opposite goals.

Again, the principle holds: energy—physical, emotional, mental—is not a vague idea. It follows patterns. It can be studied, trained and directed, and used either to harmonise or to disrupt.

## Why This Matters for Wealth and Mindset

So why bring tai chi, mudras and Dim Mak into a book about money and mindset?

Because they make something you might still be tempted to write off as 'woo-woo' impossible to ignore.

They show, in your body, what we've been saying in more abstract language:

Your inner state is not fixed.

You are not helpless in the face of every thought or emotion.

Your body is not separate from your beliefs.

You can ground yourself when volatility spikes. You can sharpen or soften your mind with a simple gesture. You can release enough tension from your posture to think clearly again. You can stay supple enough to adapt and firm enough not to snap.

The people who developed these practices were working under literal threat: fists, sticks, blades. You and I are dealing with a different kind of pressure: screens full of numbers, constant notifications, and financial risk instead of physical.

The principles, however, are the same.

Direct your intent.

Align your structure.

Let go of what you don't need.

Root your weight.

Respond rather than react.

Whether you are in a park at dawn, flowing through a tai chi form, sitting in a boardroom with your fingertips gently touching as you think, or clicking 'buy' on a trade you have planned and rehearsed, you are always manipulating energy in some direction.

The question is not whether you’re doing it. The question is whether you are doing it consciously, or whether old patterns, fears and random external forces are doing it for you.

Once you really grasp that, you stop asking, “Is this practical?” and start asking a far more powerful question:

“How can I train my state so that, under pressure, my default is not panic, but presence?”

# Chapter 11: Hypnotism & Drugs

**"Hypnosis is not mind control. It's a naturally occurring state of concentration. It's actually a means of enhancing your control over both your mind and your body."**

**Dr. David Spiegel**

The first time I really understood how strange this whole 'mind' thing is, I wasn't in a jungle with a shaman or lying in some crystal-covered temple.

I was in France, on a perfectly ordinary day, buying mushroom spawn.

I love morels and truffles. The idea of walking out into the garden, lifting some leaves and finding those things growing there felt like a small, luxurious win. So I went down the rabbit hole: what to buy, how to plant, what soil they like. Simple enough. Then I watched a documentary about how mushrooms actually work. Not the part where you fry them in butter—the part underground.

It showed this vast mycelium network: threads of fungus latching onto tree roots, linking one tree to another, moving nutrients, sharing signals when one part of the forest is under attack. Scientists call it the 'wood-wide web.'

Mushrooms... running an internet under your feet.

You don't normally think of a mushroom as intelligent. But once you see those time-lapse shots—the white threads racing through soil, wrapping around roots, passing information—you start to question that smug human assumption that we're the only clever ones in the room.

I remember sitting there thinking: if mushrooms talk to trees... what else is going on that we have no idea about?

And that's where psychedelics crept in.

The same documentary touched on psilocybin—the compound in 'magic mushrooms'—and how carefully guided psychedelic sessions were being used to treat depression, end-of-life anxiety, and addiction. Other films followed: LSD in the 1960s, modern clinical trials, people describing one well-held experience as 'ten years of therapy in a day.'

Each one poked at the same idea: change the state, change the perspective.

So my curiosity jumped again: if changing brain chemistry for a few hours can throw a door open in your mind... what exactly are we locking ourselves out of the rest of the time?

I already knew the famous story about Francis Crick, the Nobel Prize–winning co-discoverer of the DNA double helix.

Crick wasn't a monk or a mystic. He was a sharp, abrasive scientist working in post-war Cambridge. Yet there are repeated accounts of him using LSD while he was wrestling with the structure of DNA. One article even claims he admitted that LSD helped him 'visualise the spiral shape' more clearly.

We'll never know exactly how much of that eureka moment belonged to LSD and how much to years of work. But the pattern is there:

Different state.

Different angle.

Different idea.

You see the same pattern elsewhere.

Kary Mullis, who invented PCR—the technique that lets us amplify tiny pieces of DNA and underpins modern genetics and Covid testing—spoke openly about how LSD influenced his thinking. He described seeing patterns and concepts come together in ways that felt beyond his ordinary mind.

Steve Jobs called his LSD experiences 'one of the two or three most important things I have done in my life.' He believed they helped him see the world—and therefore design products—in a way that felt more intuitive, more connected, more human.

Writers, mathematicians, philosophers: Aldous Huxley, Ram Dass, and dozens of others used psychedelics as a kind of mental crowbar. Not as a permanent lifestyle, but as a way of cracking open the usual frame, taking a look around, then trying to map what they'd seen.

You can dismiss all of that as drug stories if you like. But look at the common thread:

Altered state → new perspective → breakthrough.

Nikola Tesla talked about receiving complete designs in his mind, already perfect, as if 'downloaded' from somewhere. Robert Monroe—who we'll come back to—wrote about leaving his body and travelling to other 'locales' where he seemed to access information and insight that didn't feel like normal imagination.

Whether you think those experiences are literal, symbolic, or somewhere in between, they're all pointing at the same uncomfortable possibility:

Our normal, day-to-day consciousness might not be the full picture. It might just be the narrow, survival-focused, socially acceptable slice our brain lets us use.

So the real question becomes:

If there are other modes... can we access them deliberately?

Without needing a lab, a guru, or a chemistry set?

That's the question that pulled me back to hypnosis.

Like a lot of people, my introduction to hypnosis was stage shows and TV. People forgetting their own name. People clucking like chickens. The hypnotist snapping fingers, and everyone 'falling asleep.'

I wrote it off as either acting, peer pressure, or some weird corner of human embarrassment I didn't want to touch.

Then I started reading case histories. Real ones.

There was a moment, somewhere between the YouTube clips and the research papers, when hypnosis stopped being 'maybe' in my mind and snapped into 'this is undeniably real.'

I was reading about James Esdaile, a Scottish surgeon working in India in the 1840s—long before modern anaesthetics. He reported performing hundreds of major operations, including amputations and the removal of enormous tumours, using nothing but mesmerism (what we'd now call hypnosis) for pain control. Patients lay on crude wooden tables, in conditions that would horrify a modern hospital inspector, and yet many of them remained calm and still through procedures that should have had them screaming. You can argue about Esdaile's exact numbers or methods, but you can't argue away the basic fact that human beings, under certain mental conditions, can experience radically different levels of pain from the same physical stimulus.

Fast-forward to the present, and the theatre changes from colonial surgery to MRI scanners and clinical trials, but the pattern is the same. Hypnotic analgesia—the deliberate use of hypnosis to dial down pain—has been used in documented cases of dental work, minor surgery, childbirth and chronic pain management, with patients showing brain activity changes that match their reported relief. Dr David Spiegel at Stanford has published work showing that under hypnosis, regions of the brain involved in self-monitoring and evaluation quieten down, while areas involved in focus and control light up. In other words, when someone is 'under,' they're not weak-willed or unconscious; parts of their brain are simply reconfigured so that suggestion and imagery have a more direct line into the systems that govern sensation and behaviour.

Once you know that, it becomes very hard to keep hypnosis in the 'stage trick' box. If a few carefully chosen words, delivered in a particular state, can make the difference between agony and bearable pressure on an operating table, then a few carefully chosen words, delivered in a similar state, can certainly make the difference between “I am a loser who always blows it” and “I am a competent person learning to handle success.” The mechanism is the same: change the inner picture and the identity linked to it, and the nervous system behaves differently. The proof is in blood pressure readings dropping, pain scales shifting, fMRI scans changing, and lives quietly rerouting because someone finally learned how to talk to the part of their mind that had been listening all along.

Heavy smokers who’d failed every willpower-based attempt walking into a hypnotist’s office and walking out finding the idea of smoking ‘boring,’ ‘tasteless,’ or even disgusting—because, at some core level, ‘I am a smoker’ had been swapped for ‘I am a non-smoker.’

People with phobias—flying, spiders, public speaking—who’d lived decades in avoidance, suddenly finding that the same trigger just... didn’t land the same way. The inner movie had been edited. The body didn’t panic.

Functional MRI studies at places like Stanford University (David Spiegel’s lab) have shown that in hypnotic states, brain regions involved in self-monitoring and inner criticism (parts of the default mode network) quieten down, while areas linked to focused attention and control (such as the anterior cingulate cortex) become more active—exactly what you’d expect if suggestion and imagery were gaining a more direct line to the nervous system. Clinical trials from major hospitals and medical schools in the US and Europe have found hypnosis effective for pain control, irritable bowel syndrome, anxiety, and some phobias, often matching or enhancing standard treatments.

The British Medical Association and the American Medical Association have both recognised hypnosis as a valid therapeutic tool for decades, and institutions like the Mayo Clinic include it as part of their integrative medicine programmes. In other words: serious doctors, serious scanners, and serious journals all agree that hypnosis produces measurable, reproducible changes in both brain activity and behaviour.

When I looked at it through that lens, hypnosis stopped being a circus trick and started to look like something else:

A structured, relaxed conversation with the operating system.

Here's how I now think about it.

Your conscious mind is the tiny part on the surface—the bit reading this page, analysing, doubting, planning. Hypnotists sometimes say it's like 1% of the RAM.

The subconscious is the rest.

It's the part that:

- Drives you home while you're thinking about something else.

- Reaches for your phone before you've decided to.

- Flinches at a certain tone of voice because of something that happened years ago.

- Tenses up at the number in your bank account, or at the idea of charging more.

- Automatically replays the same self-talk: “I’m unlucky,” “Something always goes wrong,” “I always blow it when it matters.”

And here’s the key: the subconscious doesn’t debate. It doesn’t sit there arguing philosophy with you.

It just runs whatever code has been installed most often and with the most emotional charge.

If you grew up with:

“Money doesn’t grow on trees.”

“Don’t get your hopes up.”

“People like us don’t get rich.”

“Be realistic.”

...then, to your subconscious, those aren’t casual comments. They’re commands.

Add to that:

“I’m always the one things break for.”

“I never quite get there.”

“I’m not that type of person.”

Say those often enough, with enough feeling, and your subconscious will quietly organise your behaviour to make them true.

So you can consciously adopt a wealth mindset—

“I am someone who builds wealth.”

“I am capable.”

“I am moving toward abundance.”

—but if the deeper part of you believes:

“I’m unlucky,”

“Good things don’t last for me,”

“Success will just bring problems.”

Then every step forward will be met with a subtle, invisible hand pulling you back.

That’s what the Law of Attraction feels like when it ‘doesn’t work.’ It’s not that visualising and aligning action is nonsense. It’s that there’s an older safety protocol whispering:

“This is too much. Back to normal. Back to safe.”

Hypnosis—done properly—is a way of sitting down with that whisper and saying:

“Okay. Tell me exactly what you’re afraid of. Let’s update this.”

The first time I really studied a hypnotic induction script, I was struck by how simple it was.

No magic words.

No crystals.

Just:

'Take a deep breath in, and as you exhale, imagine releasing any tension or stress from your body. With each breath, feel yourself becoming more relaxed and focused. Imagine a warm wave of tranquillity washing over you, soothing your mind and body. Visualise a safe space where you can let go of all distractions and embrace positivity. As you continue to relax, allow the sound of my voice to guide you deeper into a state of calmness, where you can access your inner strength and creativity. Embrace the possibility of change and transformation as you go deeper into this peaceful state.'

Strip away the soft tone, and you get a very clear process:

Slow the breath → signal 'we are safe.'

Relax the muscles → drop physical tension.

Imagine a safe place → reduce perceived threat.

Invite calm, strength, creativity → prime the subconscious for new instructions.

Once that's in place, the critical, chatty part of the mind—the bit that interrupts your visualisations with, "Who are you kidding?"—goes quiet for a while. Not gone. Just less interested in arguing.

And in that quieter space, you can finally ask questions the subconscious will answer honestly:

What limiting beliefs do I hold about myself that are affecting my success?

What fears or anxieties are preventing me from taking action?

What past experiences still impact my current mindset?

What subconscious beliefs do I need to change to stop sabotaging myself?

Asked in the middle of a busy day, those questions usually bounce off. You get surface-level answers.

Asked when your body is relaxed, your breath is slow, and your mind has stopped sprinting?

The answers float up differently.

Not as essays. Often as:

- A quick memory.

- A sentence in a parent's voice.

- A sudden tightness in the chest at a certain thought.

- A flash of you at 10 years old, being told off about money.

That's the level hypnosis lets you work at. Not arguing theory. Updating the script so that it actually runs. The science is simple: hypnosis is just distracting the logical left brain long enough to feed clear instructions to the right brain, which is always sitting there in "yes" mode.

Astral projection sits in the same neighbourhood, but it aims at a different target.

Robert Monroe didn't set out to be a mystic. He was a businessman and radio producer who started having spontaneous out-of-body experiences—lifting out of his body at night, floating near the ceiling, seeing himself asleep, then travelling to what he called "Locale I, II, III"... places that didn't feel like ordinary dreams.

Being the kind of person he was, he didn't just enjoy them or panic about them. He studied them.

He wondered: if this is real 'at all', can it be done on purpose? Can it be trained? Can other people do it?

He started experimenting with sound—what later became Hemi-Sync: binaural beats designed to guide the brain into specific frequency patterns by playing slightly different tones into each ear.

The idea was simple:

Different frequency in left ear, different in right → brain perceives a third 'beat' → brainwaves begin to follow → both hemispheres synchronise at that rhythm → consciousness shifts state.

if you play 150 Hz into your left ear and 155 Hz into your right, your ears don't hear a third sound—but your brain does. It locks onto the 5 Hz difference in the middle and starts to synchronise to that beat. In simple terms, the left side of the brain is being nudged down, the right side is being nudged up, and they meet in the middle – a convergence that can pull you into a specific state: deeper relaxation, sharper focus, or that in-between zone where the subconscious is far easier to work with.

Under those conditions, with practice, Monroe and many others reported:

- Floating above their bodies,

- Viewing distant locations,

- Encountering other consciousnesses,

- Bringing back information, insight, or at least a radically different sense of what 'I' is.

You don't have to accept every one of Monroe's conclusions to see the pattern:

Deep relaxation.

Brainwave shift.

Focused intention.

Astral projection and hypnosis work differently and aim at different outcomes.

Astral projection:

- Goal: exploration, spiritual growth, contact with 'higher' realms, non-local awareness.

- Method: deep relaxation, often audio support, intention to leave the body.

Hypnosis:

- Goal: therapeutic change—behaviour, emotion, belief; accessing and updating subconscious patterns.

- Method: deep relaxation, focused attention, suggestion often induced by a hypnotist.

But both rely on the same basic truth:

Your experience of reality changes when you change the state your mind and brain are in.

Ancient cultures understood this without EEG machines. Egyptians had 'sleep temples' where people would go to dream and receive healing or guidance. Priests used chants, rhythmic sound, incense, and imagery. They were leading people into altered states and then offering suggestions, stories, symbols.

We've given it new names:

Alpha, Theta, Delta.

Neuroplasticity.

Subconscious programming.

The territory is the same.

So where do drugs fit in this picture?

Psychedelics:

- Helps you chillax for a while.

- Disrupt the default story.

- Show you things you may not have been willing or able to see.

Hypnosis and self-hypnosis:

- Teaches you where the issue is.

- Gives you access to your subconscious.

You don't need LSD to ask:

What am I really afraid will happen if this actually works out?

What part of me believes things always go wrong, and why?

What would it feel like, in my body, to already be the person I keep visualising?

You just need:

Ten to twenty minutes,

A quieter nervous system,

And enough curiosity not to run at the first uncomfortable answer.

From there, everything else in this book becomes more than theory.

When you lie in bed at night, put on some soft theta music, breathe, relax and slip into that gentle hypnotic language; you're not just chilling out. You're opening a channel to the part of you that has been running:

"I am unlucky."

"I always fail at the last moment."

"Money is hard and dangerous."

You can thank those beliefs for trying to protect you.

Then you can offer new ones:

"I am someone for whom things increasingly work out."

"I am safe as I grow wealth and competence."

"I am a calm, competent steward of money."

"I am allowed to succeed quietly, without drama, in my own way."

And if, after that, you get up the next day and take even one small action that matches those statements, the subconscious notices.

It starts to update.

You didn't need a jungle, a guru or a heroic dose.

You needed a way to talk to yourself that went deeper than thinking.

That's what hypnotism is at its best.

Not a man on a stage with a swinging watch.

Not mind control. Just you, finally learning to sit down with your own subconscious, listen to what it's been trying to do for you, and then teach it a better way.

# Chapter 12: Self-Imposed Limits

**"The mind is the limit. As long as the mind can envision the fact that you can do something, you can do it, as long as you really believe 100 per cent."**

**Arnold Schwarzenegger**

I've already talked about wealth mindset and how "I AM" shapes your ceiling. This chapter is about turning that lens onto the invisible limits you've installed yourself. *You* start to see the real enemy up close:

Fear.

Doubt.

And the limits you quietly put on yourself... without even realising.

That's what this chapter is about.

It isn't as simple as, 'Think you are rich and it will happen.' But it is about seeing that there is a logical, structured path to adjusting how you think, feel, and show up in the world.

I'm not suggesting you run out and join a church, or let anyone with a swinging watch rewrite your head. What I am saying is this:

Once you realise how negatively conditioned you've been—by school, by governments, by media, by family patterns—you start watching films like 'The Matrix' differently.

They stop looking like fantasy and start feeling like documentaries wearing special effects.

Reality is a game. Most people wake up every day and simply react to it.

They chase money.

They chase time.

They chase approval.

They are pieces on the chessboard, not the player.

I can't overstate how important it is to build a mental version of yourself that has already won. Henry Ford's famous line—"Whether you think you can, or you think you can't, you're right"—is not just a motivational meme. It's a formula.

Because here's the uncomfortable truth that sits beneath it:

You are the one installing most of your limits.

## Self-Imposed Limits: The Invisible Bars

Start with this question:

Why can two people grow up in similar conditions, have access to the same information, read the same books, and end up with completely different lives?

Why does one break out—start businesses, invest, travel, grow—while the other stays stuck for decades?

It's not intelligence. Both might be smart.

It’s not information. The internet flattened that barrier.

It’s not luck alone.

Its limits.

More precisely, self-imposed limits.

These limits are the beliefs you accept about yourself, the ceilings you decide you’re ‘allowed’ to earn, do, or be, the lines you draw in your own mind and quietly treat as border walls.

The odd part is that most of these weren’t chosen consciously. They were suggested when you were young. They were repeated often enough that you stopped questioning them. Then your behaviour began to reflect those expectations, and your results confirmed them, so the loop tightened.

That’s where hypnosis comes back in—not as a stage trick, but as a metaphor for how we’ve all been ‘tranced’ by our environment.

In fact, some of our limiting beliefs are so deeply rooted that we don't even know we have them. We don't realise they exist and are holding us back. These triggers might not even be things we remember hearing or saying. It could be a simple phrase we heard decades ago, ' Money is hard, 'Change takes time'.

Watching hypnotist Marczell Klein literally eradicate fear and instil confidence in seconds was incredible - it made me think - I want to have that level of control on my own mind. if its possible for someone else to tap into my subconscious, I want to too!

So I started watching more of his content.

In one video, he highlighted how the mind generalises, deletes and distorts every belief, both positive and negative.

He gave an example that when he was young, kids in school said they didn't like him, so his brain generalised this simple phrase and decided it actually meant ' nobody likes him'

In fact, his mind would delete anything that contradicted this belief - if his mom said 'I love you', he dismissed this as how could someone love me if nobody likes him.

The mind distorts our beliefs and tries to alter our perception to fit into our belief system, so when girls said “I like you”, he wouldn't listen - his brain said they must be lying.

But this formula holds true in reverse, think of a beneficial belief – ‘I am the best’. The mind will generalise, delete and distort everything to reaffirm this belief.

So this is why it's so important to really look at what our beliefs are and how they are serving us. Are they helping us or holding us back?

The sooner you ' let go' of the beliefs that aren't serving you, the quicker you create space in your life for what you truly want, but if you don't truly believe that something better will come, you won't let go of what you have. The unknown can be scary, but if things don't change, they stay the same.

So, how do you discover what limiting beliefs you have?

If you heard phrases like:

"Don't get your hopes up."

"People like us don't get rich."

"Money doesn't grow on trees."

"Be realistic."

"You're not that type of person."

enough times, your subconscious didn't treat them as random noise. It took them as commands.

No swinging watch required.

You might not have consciously chosen these beliefs or even agreed with them but because you heard them enough your subconcious repeats them.

Limits Disguised as Personality.

One of the cleverest self-imposed limits is a simple sentence:

"That's just the way I am."

"I'm not a confident person."

"I'm just bad with money."

"I'm not a leader."

"I can't speak on stage."

"I'm too old for that."

"I've always been like this."

They sound like neutral descriptions, but they function as instructions.

Every time you repeat them, you're hypnotising yourself:

"I am this. I am not that."

Your subconscious listens very carefully to anything that starts with "I am." It quietly reorganises your behaviour to make those statements true.

You start avoiding situations that contradict your story. You ignore or sabotage opportunities that require you to show up as a bigger version of you. You unconsciously choose friends and environments that confirm the identity you've already accepted.

Over time, those limits disappear from view. You no longer see them as decisions you're making. You see them as 'reality'.

This is why certain hypnotic approaches can be so effective for stopping smoking, overcoming phobias, or increasing confidence. At their core, they're not inserting something alien; they're uninstalling an old programme and replacing it with something that matches what you consciously want.

The same principle applies to money, success and self-expression.

### Hypnosis as a Tool, Not a Gimmick

Think of hypnosis now as a tool.

You could, in theory, get a quick fix; it doesn't have to be a drawn-out process.

Used well, it helps you reach the deeper layers of your mind—the part that holds those automatic beliefs—without the constant interference of your over-protective, over-critical conscious mind. It lets you update the script.

Take a smoker. Rationally, they know it's bad. They may want to stop. Yet they light up anyway.

Why?

Because, subconsciously, smoking equals comfort, stress relief, a sense of identity ("I'm a smoker"), or belonging. The cigarettes are not just nicotine; they are a ritual wrapped around emotions and self-image.

Hypnosis works by separating the trigger from the response, changing what smoking means in the person's inner picture, and reinforcing a new identity: "I am a non-smoker."

When the subconscious accepts the "I am" statement, the desire to smoke often collapses.

Now swap 'smoking' for 'terrified of public speaking,' 'hopeless with money,' or 'not the entrepreneurial type.' The details change. The structure doesn't.

Someone who 'can't' speak on stage usually carries old memories of embarrassment or criticism, mental movies of stumbling and freezing, and a running commentary about being judged. Their subconscious then 'protects' them with anxiety, avoidance, and sabotage.

Change the inner movie and the “I am” statement, and the same person can stand up and use skills they already had, but couldn’t access under threat.

Wherever you’ve shrugged and said, “That’s just not me,” there’s almost always a hypnotic script running in the background.

## The Monad: What’s Beneath the Limits

If we’re going to talk about self-imposed limits, we also need to talk about what those limits are covering.

In various philosophical and spiritual traditions—Gnosticism, Neoplatonism, a few strands of mysticism—you’ll find the word Monad.

The Monad is the One, the indivisible source, the core ‘unit’ of being.

Put less academically, you could think of it as the spark of the divine inside you, the higher self, or the part of you that exists before any conditioning.

Whatever language you prefer, the idea is that beneath all the programming—beneath the fear, the doubt, the “I am not enough”—there is a part of you that knows, that is whole, and that is not limited by what teachers, parents, bosses or governments said you were allowed to be.

That inner core is creative, connected, and powerful.

If that’s even partly true, then everything we’ve covered so far—wealth mindset, faith and fortune, visualisation, the science of attraction, ancient structures and frequencies, crystals and data, technology and numbers, mind and body—is essentially about one thing:

Removing the noise so that this inner core can actually run more of the show.

Your self-imposed limits are like walls built around that core. Hypnosis and conscious change work are tools to dissolve those walls, redraw the internal map, and let more of that inner power flow into your daily life.

## How it all fits together. Look back for a second

When you adopt a wealth mindset, you're challenging the belief that scarcity is 'just how life is' and choosing to believe in abundance. If the Monad is your inner creative spark, then a wealth mindset is one way of aligning your everyday mind with that deeper nature, instead of endlessly identifying with the small, frightened version.

When we talked about faith and fortune, we defined faith as deep conviction without guarantees—trusting that your aligned actions matter. Outwardly, that might be aimed at God, the universe, or natural law. Inwardly, it's also aimed at that quiet core in you. Removing limits like "I don't deserve this" or "things never work out for me" is what creates room for genuine faith to take root.

Visualisation and 'living in the end' are ways of letting your imagination touch realities that your deeper self already knows are possible. It's like downloading pieces of a blueprint from the Monad into your conscious awareness, so you can start constructing in the physical what already exists in potential.

The science of attraction—the psychology, body language, quantum metaphors and momentum we covered—simply describes what happens when you start to live from that inner alignment. Your posture, tone and decisions shift. The external world responds.

Our detour through pyramids, frequencies, sacred geometry and crystals pointed to a larger pattern: reality itself seems structured and intelligent. If that's true, then you are not some random, disconnected accident inside it. You are a local expression of that same structuring intelligence. Limits, by that logic, are not truth—they're distortions.

Mind and body, as we saw, are one system. When you change beliefs, physiology follows. When you change physiology, new beliefs become easier to hold. That loop is the mechanism through which the Monad can either be muffled or magnified.

Seen this way, removing self-imposed limits is not about inflating your ego and pretending you're a god. It's about clearing bugs from the code so the deeper intelligence that's already there can run more cleanly.

## Evolution Through Money

Bring it back to something very concrete: money.

Making your first £100,000 is one reality. Making your first million is another. Growing to ten million or more is another again.

At each level, it isn't only the number in your account that has to change. You do.

Your identity shifts. Your decisions change. Your tolerance for risk and responsibility widens. Your emotional range expands.

Along the way, you'll bump into invisible fences:

"This is a lot of money for someone like me."

"What if I lose it all?"

"Who am I to do this?"

"My family will think I've changed."

Those are self-imposed limits.

If you don't address them, you will find ways to sabotage yourself—overspending, under-charging, freezing. You'll hit an income ceiling and call it "the market" or "the economy" when, in reality, it's your own comfort zone.

If you do address them—through honest mindset work, structured hypnotic re-programming, deliberate behavioural experiments, and deep reflection on who you think you are—then your outer life can keep expanding because your inner container can hold more.

Your Monad is already fine with expansion. That part of you doesn't know "too much" or "too big." It only knows expression.

It's your conditioned self that panics.

## Why You Snap Back to Old Feelings

You might have already experienced this.

You sit down and think about your ideal future: financial freedom, a beautiful home, the ability to travel, time to do work you actually care about, the capacity to help others. For a few seconds, maybe a minute, it feels good. Expansive. Possible.

Then your attention snaps back to the present: the job you tolerate, the stack of bills, the overdraft, the awkward email you've been avoiding, the argument you had last week.

Instead of feeling abundant, confident and joyful, you feel stressed, pressured and small.

The old feelings rush in so fast that you take them as proof. "See? This isn't for me. I was getting carried away."

What's really happening is that your subconscious is full of old programming. You have practised fear, worry and "being realistic" for years. So when you dare to imagine something bigger, your system tries to drag you back to the emotional neighbourhood it knows best.

Those feelings are not evidence about what's possible.

They are echoes. They are the emotional signature of your self-imposed limits.

So how do you start dismantling them?

You begin by exposing the script. You listen to the way you talk about yourself and your future. Every "I can't", "I always", "I never", "people like me don't..." is a line of code. Write them down. See them on paper. Ask where each one came from, when you first started saying it, and whether it's a law of physics or just a habit.

Then you deliberately flip your "I am" statements. "I'm terrible with money" becomes "I am becoming financially competent." "I hate public speaking" becomes "I am learning to speak clearly and calmly." You're not lying. You're choosing a new direction.

You use trance states wisely—self-hypnosis, guided audios, deep relaxation—to get your brain into more receptive states (alpha, theta) and feed it upgraded suggestions that match the identity you're building. You accept that repetition is non-negotiable. You spent years installing the old script; the new one deserves time.

You act 'as if' in small, grounded ways. You don't pretend to be a billionaire. You ask, "How would a slightly more confident version of me behave today?" and then you act like that person. "What choice would the wealth-building version of me make right now?" and you choose that. Behaviour shifts belief. Belief shifts behaviour. The loop upgrades itself.

And, however you choose to frame it—prayer, meditation, journaling—you make a habit of connecting with that deeper part of you, the Monad. You ask it questions: "What am I capable of that I'm not allowing?" "Where in my life am I limiting myself out of fear?" "What do I really want, beneath the noise?" Then you sit with the answers.

That quieter inner voice, the one underneath the panic and performance, is often far bolder and more generous than your surface self.

## Small Changes, Big Shifts

None of this has to happen in one dramatic breakthrough.

You don't have to destroy every limit in a single weekend.

You change one belief. You shift one habit. You drop one old story you've told yourself for too long.

Momentum begins.

You evolve. Your belief and confidence start to grow. Your awareness widens. Your faith in yourself deepens. You look back after months or years and realise that the 'you' who began this process—the one who thought "I can't," "that's not for me," "I've always been this way"—was never a fixed reality.

It was a costume. A role. A self-imposed limit wearing your name.

You are not the sum of your fears. You are not the scripts you absorbed before you were old enough to question them. You are not the ceiling other people placed over your head to keep you safe, small or manageable.

There is something deeper in you—call it Monad, higher self, soul, spark—that is bigger than any of that.

Everything in this book, including this chapter, is an invitation to strip away what you are not, reclaim what you are, stop behaving like a piece on the chessboard, and start playing the game consciously.

Imagine waking up tomorrow with amnesia and having no recollection of who you are. Now imagine you could shape the person you want to become, confident, successful, happy. You would have no limitations.

Because when you begin dismantling your self-imposed limits, you don't just get more money, more freedom, or more success. You get more of you.

# Chapter 13: Community Influence and Personal Growth

**"Setting goals is the first step in turning the invisible into the visible."**

**Tony Robbins**

There are hundreds of 'big name' motivational gurus, business coaches and life strategists out there. You know the ones.

Tony Robbins filling arenas.

Grant Cardone shouting about 10X.

Hypnotists like Marczell Klein talk about rewiring your subconscious.

Authors like Robert Kiyosaki shape how people think about money with 'Rich Dad Poor Dad'.

And that's just the surface.

Beneath that, there are probably millions of coaches, mentors, therapists, consultants and 'mindset experts' around the world. The local 'transformational' coach. The personal trainer who doubles as a life mentor. The trading educator on YouTube. The mastermind leaders are running events in cities, up and down the country.

The issue is not that these people exist. Many of them genuinely help others and have changed lives. The issue is more subtle:

How do you find the right one?

How much are you going to pay?

And more importantly: do you actually need one?

Because here's the uncomfortable truth that a lot of the self-help industry doesn't like you to look at too closely:

Once you understand the core principles, you start to realise they are all saying pretty much the same thing.

They are telling similar truths in different ways. They use different stories and metaphors. They wrap the message in different brands, personalities and price tags.

Tony Robbins talks about state, story and strategy.

Grant Cardone talks about massive action and belief.

Kiyosaki talks about assets, cash flow and mindset.

Marczell Klein talks about subconscious programming and hypnotic change.

Underneath all of it?

Belief.

Faith.

Identity.

Action.

Exactly what we've been talking about since Chapter 1.

## Same Principles, Different Packaging

Strip away the stage lights and slogans for a moment.

You walk into a Tony Robbins event, and you'll hear lines like, "Raise your standards," "Where focus goes, energy flows," "Decisions shape destiny." The delivery is high-energy, but the core ideas are about taking responsibility, directing your attention, and choosing your story.

You sit with a hypnotist like Marczell Klein, and you'll hear, "Your subconscious runs most of your life," "You can reprogram limiting beliefs," "Your identity drives your behaviour." The language is more clinical or trance-oriented, but again: it's about belief, identity and habit.

You read 'Rich Dad Poor Dad', and you're told, "The rich don't work for money; they make money work for them," "Your mindset about assets versus liabilities matters more than income," "School doesn't teach you how money really works." The angle is financial education, but under that sits the same shift: from victim to owner, from blind worker to conscious builder.

Different voices. Different stories. Different frameworks.

Zoom out and the core messages line up neatly with what we've already covered in this book: you need a wealth mindset that sees you as capable of wealth; you need faith in your future self more than blind loyalty to your past; you need to visualise your destination and 'live in the end' before it appears; you need action and consistency, not just feelings; you need to find and delete your self-imposed ceilings; you need a body–mind foundation that can carry the pressure of success.

Some people wrap those ideas in language about God or the universe. Others talk about the subconscious or vibration. Others shout about hustle and grind. Some lean on journaling and affirmations. Others emphasise NLP, hypnosis, real estate, and business.

The engine is the same.

This is crucial to grasp, because if you don't, you risk falling into a loop: thinking the next guru has the missing secret, bouncing from coach to coach, spending more time at events than actually building anything. Collecting 'vibes' instead of results.

## Why Different Influencers Pull Different People

None of this means influencers and communities are useless. Far from it. They can be incredibly powerful—if you understand what's actually happening.

When it comes to personal growth, most people need the same foundational information: take responsibility, set clear goals, know your numbers (especially with money), learn how your mind works, build habits, and choose better influences.

The content is largely universal. The way it lands is personal.

So people naturally gravitate towards different styles.

Some are jolted awake by the intense, high-energy intervention style of Tony Robbins. The sheer scale and emotion snap them out of their current frame. Others prefer a quieter, logic-heavy approach that explains everything without the fireworks. Some resonate deeply with spiritual language—God, Source, a higher power. Others want everything underpinned by science and data.

It's like a diet. One person thrives on a particular approach, not because it's the only true diet, but because it fits their preferences, lifestyle and psychology. Another tries the same thing and hates every second.

It's the same with mentors. The principles are general. The fit is individual.

If you feel drawn to a particular influencer, it's worth pausing to ask why. Is it because they help you understand and act? Or because they give you an emotional high, you've quietly become dependent on?

## The Environmental Effect

There's a reason big events and strong communities can have such a strong impact.

Imagine you've spent most of your life surrounded by people who complain, think small and shoot down every new idea. Then you walk into a room where people clap when you have a breakthrough, talk openly about investments, ideas, and growth, and treat success as normal rather than suspicious.

That contrast can be a genuine shock to your system—in a good way.

It pulls you out of your usual environment, away from the same four walls and the same limiting voices. It lets you experience belief and possibility at scale. As a human being—a social animal—you naturally mirror more of what you see around you.

That shock can interrupt old patterns, give you proof that 'people like you' can live very differently, and provide a sense of belonging to something bigger than your current circle.

There is real value in that.

The danger creeps in when you start to become addicted to the event rather than the implementation. When you begin to believe that you need that specific speaker or room to feel powerful. When you quietly outsource responsibility for your growth to the community or the coach.

No guru, no matter how charismatic, can do your push-ups for you.

They can show you the movement, get you fired up, hand you plans, and model what's possible. But the actual work—changing your habits, managing your nervous system, learning how money really works, placing trades, building products—still has to be done by you, in the ordinary hours of your own life.

## You Already Have the Core Tools

This is where everything you've read so far comes back into play.

By now, you've seen that your wealth mindset is programmable. You've seen that faith is not something handed down from a stage but an internal state you can cultivate. You've learned how visualisation and 'living in the end' can be trained. You've seen that the so-called science of attraction is rooted in your behaviour, posture and chemistry.

You've explored the idea that ancient builders might have used sound, number and geometry deliberately. You've seen how crystals, technology and the brain all operate on structure plus frequency. You've looked at your body as hardware, your mind as an operating system. You've seen through hypnosis how malleable your internal code really is. You've started to notice how many of your ceilings are self-imposed.

What does that actually mean when it comes to coaches and influencers?

It means you are not lacking secrets. You are, in all likelihood, lacking integration and consistency.

You don't automatically need a £5,000 seminar, a year-long mastermind, or weekly one-to-one calls with someone who shouts at you on Zoom. You may choose those things if the environment genuinely helps you move, if the mentorship shortens your learning curve, and if you can afford it without shackling yourself to debt.

But they are not oxygen. You will not die without them.

Most of what these people will tell you, if you condense it to its skeleton, is this: decide clearly what you want, believe it's possible and that you're worthy of it, break it into specific actions, manage your state so you don't quit when it gets hard, surround yourself with people and information that support where you're going, and keep going. The rest is flavour.

## The Risk of Idolising Influencer

If you start idolising influencers, you risk turning your power outward.

You catch yourself thinking, "When I finally make it to that event, then I'll change," or "When she notices me, then I'll believe I'm good enough." You begin copying their life instead of building your own: dressing like them, talking like them, investing exactly how they do, while conveniently forgetting your own context—your family, geography, values.

In Chapter 12, we talked about the Monad—the inner spark, the core piece of you that exists beneath conditioning. If you are constantly staring outward for authority, you stop listening to that inner signal. You override it with noise.

The irony here is that many of these influencers, if you really boil down their message, are trying to get you to claim your own power. But the way the industry is structured—upsells, tiers, continuous programmes—can accidentally push people into dependency rather than sovereignty.

You were not born to be a permanent follower. You are here to become your own authority.

## Using Influencers and Communities Wisely

So how do you navigate all this without swinging to extremes?

The first step is to check the principles behind any content you consume. When you read a book, watch a video, or sit in a room, ask yourself: Does this align with what I already know to be true from experience and from the principles in this book? Are they re-packaging the fundamentals in a way that helps me actually act?

If yes, great. If it sounds like a magic shortcut that somehow bypasses discipline, state management and responsibility, be cautious.

Next, notice how you feel 'after' engaging with a particular person or group. Do you feel more capable of directing your own life or more dependent on others to feel strong? Healthy influence leaves you with a sense of agency. Unhealthy influence leaves you feeling that you can't move without their next instruction.

Communities can and do create momentum. In a good trading group, you see others journaling, taking trades, and reviewing mistakes. In a serious business mastermind, you hear people talk about their numbers, their fears, and their wins. You feel accountable. You get feedback. All of that can accelerate your growth.

Just be careful not to let the label swallow you. You are not 'a [guru name] person' first. You are not chained to a single method for life. Use the community to learn, to build skills, to make connections—then bring what you've gained back under your own values and direction.

Your primary identity is not a brand. It's you.

## You as Your Own Best Coach

Here's the punchline.

All the tools the big names use—reframing beliefs, setting goals, visualisation, anchoring emotional states, strategic planning, habit tracking—are things you can learn and apply yourself.

This book alone has given you a framework for wealth mindset, multiple ways of looking at belief (religion, maths, quantum ideas, psychology), techniques for visualisation and living in the end, a practical understanding of frequency and momentum, concepts for aligning mind and body, insight into hypnosis and self-programming, and a way of seeing your self-imposed limits and the Monad beneath them.

You do not need anyone's permission to start using those tools.

You can create your own 'event' daily by giving yourself ten to thirty minutes of focused inner work. You can become your own hypnotist by choosing which scripts you repeat and which you drop. You can act as your own coach by asking better questions and holding yourself to higher standards.

And if you decide to bring in an external coach or join a group, treat them as a strategic partner, not a saviour.

## Bringing It Together

Influencers and communities can be powerful catalysts. They can wake you up. They can show you what's possible. They can compress time by sharing their scars and mistakes.

But the deeper message of this chapter—and really, of the entire book—is that you are not missing anything essential.

The 'thing' everyone is ultimately pointing to is not a secret they own. It is your capacity to believe; your willingness to decide; your discipline to act; your ability to regulate your state; your right to design your own life.

Some people need a firewalk with thousands of others to finally feel that. Others quietly decide it at their kitchen table on a Tuesday evening, alone.

Both are valid doors.

Either way, the end point is the same: you stop being primarily influenced, and you start becoming the influencer in your own life. Not in the social media sense, but in the literal sense.

You influence your own mind and body.

You influence how your family think about money and possibilities.

You influence the environments you walk into.

You become the person other people look at and think, "There's something different about them. I don't know what it is, but I can feel it."

And when you reach that stage, you see something very simple and very important:

You didn't get there because a guru gifted you power.

You got there because you finally accepted you had it all along—and then you did something with it.

# Chapter 14: Deeper than you Realise

**"Do not try to fix whatever comes in your life. Fix yourself in such a way that whatever comes, you will be fine."**

**Jaggi Vasudev**

If you've stayed with me from Chapter 1 to here, you'll have noticed something important creeping in around the edges of every chapter. Nothing we've talked about is as shallow as it looks on the surface.

Money, for a start, isn't just numbers shifting between accounts. Success isn't just 'work harder' and grin through the pain. Religion isn't just a set of dusty stories from a book most people haven't actually read. Trading isn't just about indicators and patterns and winning strategies. And even your thoughts aren't 'just thoughts' that appear and disappear without consequence.

You've probably also noticed that there is an entire industry built on trying to explain all of this to you, or at least claim it can. Self-help. Mindset coaching. Business coaching. Spiritual retreats. Manifestation seminars. If you've ever typed 'how to be successful' into Google or watched a motivational video on YouTube, you've seen it. It is a multi-billion-dollar machine.

Inside that machine, there are genuinely good people. There are coaches and mentors who have gone through real pain, done real work, and now want to lend a hand. There are authors who distil useful principles into accessible language. There are communities that genuinely pull people out of the mud.

But there are also rabbit holes that don't end. There are programmes that promise things they can't deliver. There are polished snake-oil salesmen who have perfected the art of telling you you're broken just enough to make you buy the 'fix'.

Everyone has a method. Everyone has a 'framework'. Everyone has 'the missing piece' they can sell if you just sign here, subscribe there, or wire the funds by Friday.

If I left it there, this chapter would just be a rant about the industry, and that's not my interest. The reason I mention this at all is that when you start connecting the dots from everything we've covered so far—from ancient constructions and symbols, to religion and Neville Goddard, to Fibonacci and fractals, to hypnosis and self-imposed limits—a much bigger picture begins to form.

And difficult as it is for the modern mind to admit, the simplest conclusion is this: there is categorically something beyond our current ability to fully understand.

That's not failure. That's a clue.

Let me take you back, not to a specific day, but to a certain phase of my life. I remember sitting in my office surrounded by books and open tabs on a browser. On the desk, neuroscience papers, a copy of Think and Grow Rich, some weird esoteric book on geometry I picked up second-hand, a Neville Goddard book with passages underlined, and a print-out of a research paper on the placebo effect.

On the screen: Bitcoin charts. Another tab with an article about Göbekli Tepe. Another with a photo of the Great

Pyramid's internal passages. Another with a snippet of a lecture about quantum entanglement. Another with a video thumbnail of a shaman talking about ayahuasca.

To an outsider, it would have looked like chaos. To be honest, it felt like chaos. But every now and then, I would feel that flicker of 'hang on a second' when two apparently unrelated things lined up. Something from an old religious teaching would echo in a modern psychology paper. A concept from trading would suddenly explain a story in the Bible. A geometric pattern carved into ancient stone would look suspiciously like a diagram in a physics book.

I realised I was doing exactly what most people who are hungry for growth end up doing: wandering through other people's maps.

You start with the motivational gurus. A Tony Robbins quote hits you hard one day: 'Setting goals is the first step in turning the invisible into the visible.' You scribble it down. Then you pick up Rich Dad Poor Dad and suddenly money looks different. It's no longer just salary and bills; it's assets and liabilities, cashflow and mindset.

You watch a video of Grant Cardone yelling about 10X and you're simultaneously inspired and mildly exhausted. You listen to a hypnotist like Marczell Klein talk about subconscious programmes and identity, and you hear echoes of what Neville Goddard was saying decades earlier about feeling, assumption, and imagination.

Different voices. Different brands. Different stage backdrops. But as you listen carefully, you realise they're all circling the same simple truths:

You become what you repeatedly think and feel.

You act out the story you secretly believe about yourself.

You get what you tolerate.

You are shaped by your environment and by what you allow into your mind.

We've been hammering those themes throughout this book, from a trader's angle, from a psychological angle, from a slightly heretical religious angle, and now from a more openly 'what on earth is this place?' angle.

At some point, you have to stop asking, "Who has the secret?" and start asking, "What are all of these people, traditions, and structures pointing at?"

I want you to imagine this for a moment.

You're standing in front of the Great Pyramid of Giza. You've seen the photos, but the real thing is different. It's far bigger, far stranger, and somehow it feels more alive than you expected. You run your hand across the stone. You tilt your head back and try to see the top against the sky. People mill around. Tourists pose for photos.

Officially, you've been told this is a tomb. A big, impressive, incredibly precise tomb for a single man. You've been told about ramps, chisels, ropes, and endless lines of human labour costing twenty-odd years of effort.

But you've also stood inside, or at least seen photos and videos of the inner chambers—those silent granitic spaces with no hieroglyphs bragging about the pharaoh, no Book of the Dead carved into the walls. You've read about the mathematical relationships hidden in its dimensions, about how it's aligned more precisely to true north than many modern buildings, about its acoustic properties at specific frequencies.

And because you've read this book, you're now thinking about Fibonacci ratios, about F-sharp, about standing waves, about sound experiments with sand forming geometric patterns. You're thinking about Tesla and his obsessions. You're thinking about how crystals can hold data and how quartz vibrates.

The pyramid stops being just a tourist attraction. It becomes a question mark. A statement in stone that says, "There is something you're missing."

Now take a mental flight over to the Amazon.

You're in a simple wooden structure with a group of people, in the middle of a dense jungle. The air is thick and alive with sounds. A shaman sits opposite you, holding a cup filled with a foul-tasting brew. Ayahuasca. You've heard the stories. Visions. Spirits. Patterns. Trauma surfacing and healing. People meeting their fear, their ego, their past, their 'guides'.

You drink. Time does strange things. The world melts and reforms. At some point, you are no longer sure what is 'in your head' and what is 'out there'. You see geometry. You feel a presence. You feel connected to something that makes the usual daily worries look like ants on a motorway.

When you come back—and you do come back—you're holding fragments in your hands. Insights. New stories. A different way of feeling your own life. You might put it down to brain chemistry. The shaman might put it down to plant spirits. But either way, you have experienced a layer of reality that was there all along, which you'd never have accessed sitting on your sofa at home.

Fly again.

Now you're in a modern laboratory, watching a simulation of the double-slit experiment. You see dots forming an interference pattern on the screen when nobody checks 'which slit' the photons go through. You see that pattern collapse into something more classical the moment measurement is introduced. You listen to a physicist explain wave-particle duality, the collapse of the wavefunction, and the unsettling implications of observation affecting outcome.

Somewhere in the back of your mind, you remember religious lines about 'the Word', about 'let there be light', about faith the size of a mustard seed moving mountains. You remember Neville saying that feeling is the secret. You remember reading about the placebo effect—people healing because they believed in sugar pills. You remember sports psychology studies where mental rehearsal activates the same pathways as physical practice.

Then, in the same afternoon, you bump into a book about the Akashic Records, describing an invisible archive of every thought, word, and deed, and another about the subconscious mind as your inner data bank. Two completely different vocabularies, but both describing something like an information field that shapes your experience.

And you start to feel that same tug again: “Something is going on here.”

The phrase “Akashic Records” might sound a bit out there if you’re coming from a plumbing background, or a police officer’s, or a traditional religious one. But step back from the label for a moment and just look at the structure of the idea.

The Akashic Records are described as a kind of non-physical library. A compendium of all thoughts, experiences, and actions, across all souls, across all time. You don’t access it with a library card; you access it through altered states—deep meditation, trance, occasionally under the influence of certain substances, or through what some call ‘downloads’ or intuition.

Now park that and think about the subconscious mind.

We’ve already discussed how your subconscious stores your memories, your emotional charges, and your learned beliefs. It’s the layer of mind that runs programmes while you’re busy doing something else. You don’t have to ‘decide’ how to walk or drive once you’ve learned or read a sentence. Programmes kick in automatically.

The subconscious holds your personal archive. The Akashic Records, if they exist, are described as a kind of collective archive.

Some people believe that when you learn to work with your subconscious—through hypnosis, visualisation, deep introspection—you’re sometimes doing more than cleaning up your old patterns. You are, in those quiet, focused, altered moments, putting your hand briefly into a deeper field of information.

Is that literally true? I don't know. And I'm not pretending to.

But on a practical level, it almost doesn't matter whether the Akashic Records are a literal external field or a poetic way of describing how the deeper layers of the mind connect with each other and with reality.

What matters is that:

You are more than your conscious chatter.

You have access to more than you usually tap into.

You can go inward and bring back something that didn't seem to be there before.

And that fits perfectly with what Neville talked about when he described states and imagination. It fits with what various religions talk about when they refer to 'revelation', 'guidance', 'inspiration'. It fits with what creatives describe when they talk about ideas 'arriving' fully formed.

It also fits, quite neatly, with how modern science is being forced to admit that information and observation are not neutral side notes in the workings of the universe. They are woven into the fabric itself.

Now let's slide from brain science and metaphysics back to stone and symbol.

If you look carefully, you'll see that almost every major civilisation has used an overlapping set of tools to try to express and access the deeper layers of reality:

Geometry.

Number.

Sound.

Ritual.

Story.

Freemasons use the compass, the square, and the compass as teaching symbols about building a better self. Cathedrals in Europe are built according to precise ratios; they channel light and sound in deliberate ways. Temples in Asia align with celestial events. Mosques, churches, synagogues, shrines—they're all built to create a particular feeling in the human who steps inside.

In the Vatican, beneath the surface of museums and tourist routes, there are archives that ordinary people will never see. At one time, the Library of Alexandria housed knowledge from across the known world, only to be destroyed—whether in a single fire or several events, we may never know. Ancient inscriptions that might have told us more about how people saw the universe have been worn away, removed, or simply never translated.

Even if you strip away every wild theory about hidden orders and secret elites, these simple facts remain:

There has always been a sense that some knowledge is not for everyone. There has always been an understanding, at least in some circles, that reality has layers. And there has always been an urge to encode important truths in forms that can survive time: stone, story, symbol.

Why? Because images and shapes can smuggle meaning past the rational mind and lodge in the deeper layers.

You know this already.

When a trader first looked at charts, they were just lines and candles. After a while, they started to see structures. They could look at something and say “accumulation”, “distribution”, “Liquidity grab”, “UTAD”, “spring”. A rectangle on a chart became an Auction Area in their mind. A tiny shift in volume suddenly told a story about buyers and sellers and Composite Man.

The picture and the pattern became a language.

The same is true elsewhere. The Eye of Providence on a dollar bill, the arrangement of pillars at a temple entrance, the three steps here, the twelve signs there, the repeated use of 7, 3, 33... maybe not everything is a code, but let’s not pretend it’s all random either.

Once you accept that symbols, ratios, and shapes have been used for centuries to carry meaning, you can no longer scoff at the idea that the ancients might have known a few things we have forgotten, or at least buried under concrete and smartphones.

At this point, a reasonable question is: “Alright, but what do I do with all this?”

I want to be very clear about something. It is not your job—or mine—to come up with a grand unified theory of everything before you allow yourself to build a life you want. You don’t need to prove that the Akashic Records exist in a lab to benefit from journaling, meditation, or hypnosis. You don't need to decode every carving on every temple to use geometry and structure in your own thinking. You don't need to confirm every story about Tesla or Coral Castle to take your own flashes of insight seriously.

Our job here is simpler, and, in my view, more honest.

First, we can acknowledge that reality goes deeper than we were led to believe in school. The picture we were given—dead matter, random evolution, religion as fairy tales, mind as a by-product of brain activity—is at best incomplete.

Second, we can stay open without letting our brains leak out of our ears. That means we don't have to believe every story we hear. We also don't have to dismiss every experience, every pattern, every symbol that doesn't fit neatly into a textbook.

Third, we can decide to use what is useful and visible now.

You don't need a PhD in metaphysics to see that:

Your thoughts and beliefs shape what you notice and how you act.

Your body and mind are in constant conversation.

Your emotional state colours your decisions, your timing, and your risk tolerance.

Habitual visualisation and feeling change your brain and your behaviour.

Rehearsing failure all day produces more of the same. Rehearsing success, from a place of true belief, changes how you move in the world.

Self-talk, like "I am this" and "I am not that", quietly locks you into or frees you from certain outcomes.

You have seen, throughout this book, that trading is not just clicking buttons. It is interacting with an algorithmic, pattern-seeking market. You've seen that the way you show up mentally and physically before, during, and after a trade is as important as any indicator. You've seen that your internal settings around worth, wealth, and possibility will cap or expand your financial results—no matter how clever your system is.

You have also seen, I hope, that everything from Fibonacci levels on a chart to spirals in a galaxy, from a quiet cathedral to a humming jungle hut, from a fMRI scan to a Freemason tracing board, is whispering the same kind of things:

There are patterns.

There is structure.

There is depth.

There is more.

The mistake is to think you must understand all of it before you are allowed to move.

I remember one afternoon, sitting at that same desk I mentioned earlier, thinking about all this. Pyramids, photons, Tesla, shamans, Vatican vaults, Coral Castle, hypnosis scripts, Livermore quotes. It was tempting, in that moment, to push everything to one extreme or the other.

On one hand, you can say, "It's all nonsense. The world is rocks and chemicals and empty space, and anything beyond that is people seeing what they want to see."

On the other hand, you can tumble head-first into every theory that takes your fancy. Every symbol is a code. Every odd number is a message. Every coincidence is deliberate. You start spending more time chasing secrets than living your life.

Both extremes are traps.

What I landed on, and what I'm offering you, is a middle ground.

We can afford to be humble enough to say, "I don't know exactly what's going on here"—and at the same time practical enough to say, "but I can see enough to live differently."

You have enough information already to:

Treat your thoughts and emotions as levers, not accidents.

Treat your body as hardware that needs maintenance, not an afterthought.

Treat your beliefs as scripts you can edit, not verdicts carved into stone.

Treat your imagination as a tool for rehearsal and alignment, not a toy to be shut down at childhood's end.

Treat your environment—people, spaces, information—as part of your signal.

Treat your setbacks as feedback from a complex, responsive system, not punishments from a cruel universe.

You don't need a secret handshake for that. You don't need approval from a guru for that. You don't need to join a lodge, a church, a cult, or an online mastermind for that.

If you choose to engage with communities, mentors, or teachings from any of these worlds—modern coaching, ancient wisdom, religious frameworks, business education—do it consciously. Recognise that most of them are simply different ways of pointing toward the same underlying principles:

YOU are part of something bigger.

YOU can change how you interact with it.

YOU do not have to stay inside the limits you inherited.

Everything else is detail.

In the end, this chapter is titled 'Deeper Than You Realise' for one simple reason.

Most people graze on the surface of life. They get up, go to work, worry, scroll, sleep, repeat. They never question the story they've been handed. If they feel that itch that there must be more, they quickly smother it under distraction or a few motivational quotes.

You've chosen to do something different just by reading this far.

You've looked at money, markets, mindset, faith, vibration, geometry, hypnosis, and self-imposed limits from angles most people never bother to explore. You've peeked behind curtains—religious, scientific, historical—and seen that none of these domains has all the answers on its own.

That can be disorienting at first. But it's also an invitation.

The invitation is not to become a full-time mystic or a professional conspiracy theorist. The invitation is not to throw away your brain and drift into blind belief. The invitation is simply this:

Acknowledge that life runs deeper than the story you were given.

Stay curious.

Use what you do know.

And keep expanding your capacity to live on purpose.

You will never know every corner of this universe. Neither will I. Neither will any guru, priest, scientist, or shaman.

But you can know enough to stop living as if you are powerless.

You can know enough to stop being a piece on the chessboard, and start being the player moving the pieces.

You can know enough to wake up each day with the quiet, grounded realisation:

"I am part of something vast. I may not understand it all. I don't need to.

But I can choose my mindset, I can direct my focus, I can honour my body, I can refine my signal,

and I can walk forward as if my life matters—because it does."

That is more than most people ever do.

And that, for me, is where deeper understanding really begins.

# Chapter 15: CIA

**"Classified information includes much more than the actual 'secrets' acquired. It includes how they were acquired and the process by which related analyses were made."**

**Will Hurd**

This chapter might feel a little long, but I really wanted to emphasise the fact that the CIA take this as seriously as they do.

The first time I realised just how far down the rabbit hole this all went, I wasn't in a jungle with a shaman, or sitting in an ancient temple. I was sat at a desk, in front of a very unspiritual thing:

A declassified government PDF.

The file itself looked utterly boring.

Black-and-white. Courier font. Stamps across the top: "Approved For Release". Dates. Reference numbers. The sort of thing you'd expect to find about troop movements or budget allocations, not about the nature of consciousness.

Across the top of the first page, it read:

"DEPARTMENT OF THE ARMY

UNITED STATES ARMY OPERATIONAL GROUP

US ARMY INTELLIGENCE AND SECURITY COMMAND

Fort Meade, Maryland

SUBJECT: Analysis and Assessment of Gateway Process"

Not a YouTube thumbnail.

Not a self-help book.

Not some anonymous blog.

An internal document from military intelligence, written in the early 1980s.

I started reading.

And the deeper I went, the more familiar—and at the same time, disturbing—it felt. Because what this dry, technical report was talking about was not new. I'd seen it in places most people would call 'woo-woo': meditation, hypnosis, visualisation, Neville Goddard lectures, shamanic stories, even in the way some religious texts hint at altered states.

But here it was in black and white, being dissected with the same cold language you'd use to describe a radar system:

Brainwave entrainment.

Binaural beats.

Hemi-Sync tapes.

Out-of-body experience.

Holographic universe.

Consciousness interacting with a field.

Let me walk you through what I saw, because it matters for everything we've built in this book so far.

It isn't just one paper, there's a whole bunch of them. From whitepapers to patents.

The report was written as an attempt to analyse the Monroe Institute's 'Gateway Experience'. If you're not familiar with Monroe, he was one of the early pioneers of using sound—specifically binaural beats—to shift people into altered states of consciousness.

The basic idea was simple:

Play slightly different tones in each ear → the brain perceives a beat frequency equal to the difference between them → the brain starts to 'follow' that frequency (frequency following response) → both hemispheres sync up at that desired rhythm.

Mix that with guided relaxation, visualisation, and breathing, and you've got a recipe for shifting people out of normal waking consciousness into something deeper.

The Army wanted to know:

Is there anything real here?

Can this be mapped, measured, understood?

And, very quietly behind the scenes: could it be used?

As I read, I couldn't help smiling grimly. Out there in the public, people made fun of meditation, laughed at 'mind over matter', mocked the idea of astral projection and remote viewing.

Meanwhile, in this report, the author was patiently explaining:

- How the left and right hemispheres of the brain work.

- How you can use sound to synchronise them.

- How these synchronised states correlate with meditative and trance states.

It was like reading a scientific commentary on the earlier chapters of this book:

Yes, the brain can be steered into different states with rhythm and frequency.

Yes, imagery and emotion in those states can have profound effects.

Yes, there's a lot more to consciousness than everyday thinking.

The report talked about the Frequency Following Response—the documented phenomenon where the brain locks onto a repetitive stimulus and starts to mirror it. That's exactly the principle behind certain meditation tracks, hypnosis audios, and even some types of religious chanting.

It went on to describe in careful, almost clinical detail how participants in the Gateway process were trained to:

- Relax the body deeply.

- Focus the mind.

- Visualise energy moving through and around them.

- Reach states where they could observe themselves from outside their bodies.

- Potentially perceive information at a distance.

Not once did the author call it nonsense.

He analysed it.

He compared it to hypnosis, transcendental meditation, and biofeedback. He referenced neurology and physics. He talked about states, frequencies, and patterns.

I remember sitting back in my chair, PDF open, thinking:

"If this is all woo-woo, why is the US Army Intelligence and Security Command writing multi-page reports about it?"

Around the same period, I had been digging into other threads as well.

One was the story you might have heard about Nikola Tesla.

Tesla died in a New York hotel room in 1943. Officially, he died alone, in debt, his genius mostly unrecognised in the mainstream. Shortly after his death, representatives from the US government—historically, the Office of Alien Property Custodian and the FBI—moved in and seized his papers and effects.

You can find references in declassified memos and secondary sources to those documents being reviewed, some labelled as unimportant, others flagged and routed somewhere else. There are claims—some more solid, some pure speculation—that certain notes and plans vanished into classified projects.

What we do know is that:

- Tesla was decades ahead on AC power, wireless transmission concepts, and resonant energy ideas.

- He talked openly about receiving complete designs as visions, almost like someone downloading CAD files into his mind.

- After his death, his work was treated not as the ramblings of a crank, but as potentially sensitive material.

Again, the pattern:

In public, you can dismiss Tesla as eccentric.

Behind closed doors, you treat his notes like they might contain strategic gold.

If you keep following that pattern, you run into other unsettling corridors.

You find programmes like MKULTRA—a series of CIA projects in the 1950s and 60s aimed at exploring mind control, interrogation techniques, and the effects of drugs and sensory deprivation on consciousness. Much of what was done under that umbrella was unethical, to put it kindly. Some of it veered into the monstrous.

You read testimonies and declassified summaries of experiments with LSD, hypnosis, sleep deprivation, subliminal messages, and attempts to break and rebuild personalities. You see that the idea of 'programming the mind' was not just the domain of stage hypnotists and paperback authors. Governments poured money and time into it.

Not because they believed in magic wands.

Because they believed the human mind could be steered, hacked, exploited.

Then, in other documents, you stumble on odd little phrases that feel like they've slipped in from a science fiction novel:

- 'Remote viewing.'

- 'Out-of-body experience.'

- 'Psychotronic research.'

- 'Altered states of consciousness.'

- 'Walking through a wall'—appears in reports and testimonies as the alleged outcome of some extreme altered state experiments, or at least as an idea entertained enough to warrant testing.

Some of these claims are clearly exaggerated or unproven. You don't have to buy every story. But the fact that they appear at all, in the context they do, is telling.

When intelligence agencies formally commission and archive work under titles like 'Analysis and Assessment of Gateway Process' and 'Analysis and Assessment of Out-of-Body Techniques', you can safely conclude one thing:

This subject was, and is, taken seriously behind the curtain.

Go back to that Gateway report.

Further into the document, the language shifts from strictly neurological to something far more ambitious.

The author starts talking about the holographic model of reality. Drawing on physicists like David Bohm and neurologists like Karl Pribram, he outlines a view where:

- Reality can be understood as a vast interference pattern of energy.

- Everything in that pattern contains information about the whole (just like a hologram).

- The brain doesn't create consciousness but filters and decodes this field into the specific 3D experience you and I call 'the world'.

On one page there's a simple diagram: a cube labelled 'Consciousness Energy Grid,' with the left and right hemispheres interacting: one reducing input to verbal symbols and concepts, the other handling spatial, pattern-based information.

Elsewhere, he draws a torus—a donut-shaped field—representing a universal hologram, with energy cycling from 'the Absolute' through time-space and back again.

Read that slowly:

In an official, approved-for-release document, an Army analyst is describing the universe as a hologram, consciousness as energy interacting with that hologram, and certain sound-based practices as ways to shift the way consciousness moves within that field.

Now bring to mind some of the threads we've woven through this book:

- The Monad inside you—the spark beyond conditioning.

- The Akashic Records as a field of information rather than a library on a cloud.

- The idea that your "I AM" statements are not just casual phrases; they're commands affecting how you interface with reality.

- The notion that religion, at its core, might be an instruction manual for aligning with deeper laws.

- The way sacred geometry, sound, and geometry in stone seem to encode an understanding of frequency and pattern.

Suddenly, what once looked like disconnected fringe ideas start lining up with someone in a military office, in 1983, trying to build a model that can hold:

- Brain science.

- Quantum theory.

- Holographic universes.

- Meditation.

- Out-of-body experiences.

You couldn't write this as fiction without people calling it over the top.

Yet here it is, in a government archive, stamped and filed.

Let's be honest: people walk away from this kind of material in very different directions.

Some go, "Ah, see? Proof of everything I already believe."

Others say, "No, that's just experiments and speculation; it doesn't prove anything."

What interests me, in the context of this book, is not making you choose a particular camp.

What interests me is your willingness to look at the pattern.

On one level, you have:

- Tesla giving interviews about receiving inventions as clear inner visions.

- Shamans guiding people into visionary states with plants and drumming.

- Neville Goddard telling audiences that imagination creates reality if you live in the end state.

- Religious mystics across traditions describe ecstatic states, visions, mergings with something vast.

- Traders visualising successful trades and notice performance improvements.

On another level, you have:

- Intelligence agencies exploring hypnosis, drugs, and sensory manipulation to control or liberate consciousness.

- The Army commissioning formal evaluations of 'gateway' techniques to expand perception.

- Declassified documents discussing remote viewing and non-local awareness.

- Reports—some more credible than others—of people experiencing OBEs that seemed to allow them to perceive distant places or events.

Overlay those maps, and a shared contour appears:

Consciousness is not confined to normal waking perception.

It can be shifted, trained, distorted, and expanded.

Sound, image, belief, and expectation are levers.

Some of these levers have been used openly in religion and self-help.

Others have been pulled in secret, in labs and bunkers.

That alone should be enough to make you pause before dismissing any of this as mere fantasy.

There's a phrase that appears in some of the remote viewing and Gateway-related discussions that I find strangely grounding: 'operational usefulness'.

Behind all the theory, all the esoteric language, the question investigators kept circling back to was simple:

"Can this be used for anything?"

That is, can non-ordinary states of consciousness produce:

- Information that is verifiable?

- Changes in behaviour that are measurable?

- Advantages in intelligence, communication, healing, or performance?

You and I don't need to decide, in this chapter, to what extent those experiments 'worked'. The public record is patchy. Some programs were cancelled. Some were buried. Some participants insist they saw astonishing results; others saw noise and false hits.

What I take from it is this:

The idea that consciousness can reach beyond the skull—into space, into time, into information fields—was considered plausible enough, often enough, for serious people to test it, refine it, and in some cases, quietly keep going even when the front-facing program was shut down.

Meanwhile, you have a world full of ordinary people who've had experiences they can't quite explain:

The dream that later plays out in real life.

The sudden knowing that a loved one is in trouble, right before the phone rings.

The strong sense to avoid a certain road or decision that later proves wise.

The visions or messages in deep meditation or under plant medicine that offer information the person might not, in theory, have had access to.

Most of these experiences get shoved into a mental cupboard labelled 'weird stuff' and left there.

But put them alongside:

- Shamans' maps of spirit realms.

- Religious descriptions of revelation and prophecy.

- Neville's stories about clients manifesting specific outcomes through feeling.

- The Gateway report's diagrams of consciousness interacting with a holographic field.

...and you begin to see why some people, in some offices, thought it was at least worth running a few classified experiments.

Now, let's bring this back down to earth—your earth.

Because up to this point, you might be thinking, "Interesting, but I don't work for the CIA, I'm not Tesla reincarnated, and I'm not planning to walk through any walls. What has this got to do with my life, my wealth, my mindset?"

Everything.

Not because you're going to start remote viewing for a living, or sign up for some questionable mind-control experiment, but because the same underlying principle applies:

Your consciousness is more malleable and more powerful than you were told.

That's it. That's the core message.

If:

- Militaries have tried to shape it for strategic gain.

- Marketers and propagandists shape it every day to sell you things or influence your votes.

- Shamans shape it for healing and insight.

- Religious traditions shape it for devotion and moral formation.

- Self-help speakers shape it (for better or worse) to motivate and direct.

- And you, unconsciously, shape it all day with your own self-talk, images, and emotional rehearsals...

...then it's time you stopped treating your mind as a passive recipient.

You already know, from earlier chapters, that:

- Your beliefs form the ceiling on what you allow yourself to attempt.

- Your identity ("I am..." statements) acts like a script you follow.

- Visualising your future, with emotion, changes your brain and behaviour.

- Small, daily choices change your chemistry, posture, opportunities.

- The universe (or market, or people around you) tends to respond to your consistent signal more than to your sporadic wishes.

The added layer from these declassified documents is not that they give you a magical new technique.

It's that they offer institutional acknowledgement that:

Consciousness is not a simple by-product of grey matter.

It interacts with fields, patterns, perhaps dimensions we don't fully grasp.

And techniques that alter your state—sound, suggestion, focus—have real effects.

If you put that together with everything else we've covered—sacred geometry, ancient alignments, religious metaphors, neural plasticity, mind–body feedback—you arrive at a simple but profound conclusion:

There almost certainly is something real above our level of current understanding.

Call it the field.

Call it the Absolute.

Call it God, source, the hologram, the Akashic, the Monad, the matrix.

The label is less important than the stance you take.

So, where does that leave you as you read this, possibly with a coffee nearby, a chart open in another tab, your own life's problems wandering around in the background like patient dogs?

It leaves you with a choice.

You can look at all of this—Tesla's seized papers, CIA mind-control projects, Army analyses of out-of-body experiences, declassified references to consciousness as a holographic field—and shrug.

'Interesting. Strange. Back to 'real life'.'

Or you can let it crack your certainty just enough to let something new in.

Not blind belief.

Not paranoia.

Just a quiet, grounded admission:

'I've been living as if reality is much simpler than it actually is.

I've been treating my mind as if it's fixed, my body as if it's separate, my thoughts as if they're private and powerless.

But what if none of that is true?

What if my inner world and outer world are entangled in ways I've barely begun to explore?'

From that question, everything you've read so far in this book takes on a different weight.

Wealth mindset is no longer a nice idea; it's an act of aligning with deeper laws of expectation and probability.

Visualisation stops being childish and becomes a way of instructing your nervous system and, maybe, interfacing with a larger field.

Affirmations and "I AM" statements stop being cheesy and become explicit commands to your own operating system.

Choosing the spaces you enter, the symbols you surround yourself with, and the sounds you expose yourself to, stops being aesthetic and becomes energetic hygiene.

Rejecting self-imposed limits becomes more than productivity advice; it becomes a reclaiming of your portion of whatever this larger intelligence is.

You don't need to levitate.

You don't need to read classified documents for breakfast.

You don't need to walk through walls.

You just need to stop walking through your own life as if you're powerless.

One of the temptations, when you lift the curtain this far, is to get stuck staring at the machinery.

You could spend the next ten years chasing every declassified file, every leaked memo, every story about underground facilities with names like HAYSTACK or projects with codenames that sound like B-movie scripts. You could lose months in forums arguing about what Tesla did or didn't invent, what the CIA can or can't do, whether someone, somewhere, has cracked astral travel as a reliable tool.

If that's your hobby, fine. But that is not what this book is for.

This book is for something much closer to home.

It's for the version of you that wakes up tomorrow.

It's for the version of you that decides what to believe about money, about self, about possibility.

The version that either:

- Nods politely at all of this and carries on living in the same mental prison, or

- Takes it as sufficient proof—not absolute, scientific proof, but enough—that there is something bigger going on, and that you are allowed to start exploring your role in it.

You and I might never know exactly what happened in every underground lab, what was written on every page of Tesla's missing notebooks, or how far certain experiments with mind and matter have really gone.

But we don't need that level of detail to live differently.

We only need to see the outline.

And the outline, at this point, is hard to ignore:

Your mind is not small.

Your beliefs are not trivial.

Your inner images are not toys.

Your state is not incidental.

Your life is not disconnected from the deeper patterns woven through pyramids, cathedrals, star maps, sacred diagrams, declassified theories, and the strange, persistent feeling that 'there must be more than this'.

There is.

You may not be able to name it fully.

You may not be able to prove it in a lab tomorrow.

But you can feel around its edges.

You can cooperate with its hints.

You can, at the very least, stop pretending it's all just random noise.

Behind the curtain, the people with budgets and clearances have been poking at this for decades.

On the surface, you have been told to laugh at it.

In these pages, I'm asking you to do something very simple, and very radical:

Stop laughing.

Start listening.

And then, step by step, start living as if your thoughts, your state, and your decisions carry far more weight in the fabric of things than you were ever told in school.

The systems behind the curtain clearly believe that might be true.

Maybe it's time you did too.

# Chapter 16: Sacred Geometry

**"Learn how to see. Realise that everything connects to everything else."**

Leonardo da Vinci

It usually starts the same way.

You walk into a building, or you stand in front of a monument, and before you have time to think about it, something in you goes quiet. Your shoulders drop. Your breathing changes. Your eyes instinctively follow certain lines and curves. You don't know why. You just feel it.

Maybe it's a cathedral in Europe, a mosque in Istanbul, a temple in India, a pyramid in Egypt, or even a simple stone circle standing silently in a field. The age, the culture, the religion might change, but the feeling doesn't. There is something about the shape of the place that speaks a language your conscious mind doesn't.

When I first started to connect the dots between mindset, religion, ancient structures and science, I didn't expect geometry to show up as one of the main characters. Numbers, yes. Probability, definitely. But shapes? Ratios? Patterns carved in stone thousands of years ago?

Yet the more I read, the more I travelled, the more I stared at charts and buildings and spirals on shells, the more one idea kept knocking at the door:

The universe is not just made of 'things'. It's made of relationships. Proportions. Shapes.

And when those shapes and proportions repeat often enough, in enough places, for long enough, they earn a name: Sacred geometry.

Let's go back to the foot of the Great Pyramid. We've already stood here in earlier chapters, blinking in the harsh light, picturing the logistics of hauling millions of stone blocks into place. This time, imagine you're not a tourist with a camera, but someone with a different job: to pay attention to where this thing is pointing.

You've got a simple compass in your hand. You walk around the base and discover that its sides are almost perfectly aligned with true north, south, east and west. Not roughly. Not 'good enough.' Astonishingly precise—especially when you remember that, according to the school version of history, this civilisation didn't have access to our modern instruments.

The Great Pyramid was not thrown together as a crude tomb. The way it's built and what it's built from scream 'different purpose.' It's a stone mountain, stacked 481 feet high, covering over 13 acres. Yet its sides are aligned to true north with an accuracy that modern engineering would struggle to match. Within about 1/15 of a degree. The base is level to within three-quarters of an inch. Today, you'd use lasers to get that right.

Each side is over 755 feet long, and the difference between any two sides is within about two inches—that's 99.98% accurate. And it doesn't even have four flat faces. It has eight. Each side is slightly concave, a detail you only really see from above or in the equinox shadows. Those hidden angles are perfect, too.

Now step back further in your mind, far enough that you can see the whole Giza plateau from above. The three main pyramids are not scattered randomly. Their arrangement suggests a pattern. Some researchers argue that they mirror the belt stars of Orion. Others debate how exact that match really is, but even if you ignore the Orion theory completely, it's hard to look at the layout and say, "Ah yes, purely casual."

Something is being expressed here:

Stone speaking to sky. Earth answering heaven. Geometry on the ground reflects geometry above.

Now move your mental camera to night. The desert cools. The sky darkens into ink. Millions of stars appear—far more than most of us ever see through city light pollution. The Milky Way arcs overhead like a river of light. Constellations rise and fall. The same patterns return, season after season.

Over the years, if you're an ancient observer with no television and plenty of time, you begin to notice relationships. Certain stars appear at specific times of year, signalling flooding, planting, and harvesting. The Sun rises and sets at particular points on the horizon on solstices and equinoxes. If you're curious and patient, you start marking things: shadows on the ground, lines in the sand, stones in circles.

Gradually, a picture forms in your mind:

The heavens are not chaotic. They move according to rules. Angles. Intervals.

If you think like the people who built the pyramids, another thought might follow:

'If the heavens move according to geometry... perhaps our most important structures should too.'

So you build with intention. Not just for shelter. Not just for storage. But as statements in stone.

## Cathedrals, Proportion and the Body

Now fast-forward several centuries and shift your point of view inside a grand cathedral.

The air is cool and faintly scented with stone, dust and incense. Light pours through stained glass windows in deep colours, falling onto the floor in pools of red, blue and gold. Your footsteps echo. Without meaning to, your head tilts back. The ceiling stretches impossibly high above you in ribs and arches, like the inside of some gigantic, petrified tree.

You don't know the architectural plan. You don't know the ratios the builders used. You might not even know the religious stories in the statues and windows around you. Yet you feel something.

Part of that is the history and expectation you carry with you. But part of it is purely physical. Your body is responding to proportion.

Tall vertical lines lift your gaze. Repeating arches create rhythm in space. Columns divide the interior into modules—bays that echo each other. The width of the nave, the height of the ceiling, the spacing of the windows: these are not accidents. Many cathedrals and temples are built using consistent ratios—halves, thirds, fifths—and in some cases the golden ratio, the same mysterious proportion that shows up in sunflowers and seashells.

This is what people mean when they talk about sacred geometry: the deliberate use of shape, proportion and pattern to express something about the way reality is put together.

'Sacred' here doesn't have to mean mystical or irrational. It simply means 'set apart' or 'dedicated'. These are not just useful forms; they are meaningful forms.

## Fibonacci, Pi and Spirals Everywhere

We touched on Fibonacci numbers earlier. Let's go a bit deeper, without getting lost in equations.

Start with zero and one. Add them: 0 + 1 = 1.

Add the last two numbers again: 1 + 1 = 2.

Then 1 + 2 = 3.

2 + 3 = 5.

3 + 5 = 8.

5 + 8 = 13.

8 + 13 = 21.

Keep going like that, and you generate the Fibonacci sequence:

0, 1, 1, 2, 3, 5, 8, 13, 21...

Very basic. Very mechanical. There's nothing magical about the adding itself.

But now take two neighbouring numbers from further along the sequence and divide the bigger by the smaller: 13 ÷ 8, 21 ÷ 13, 34 ÷ 21, and so on. As you move out, those ratios converge on an irrational number approximately 1.618.

That ratio pops up everywhere.

You see it in the spirals of nautilus shells.

In the pattern of seeds in a sunflower head.

In the swirling arms of some galaxies.

In the branching of many trees.

In the proportions of classical buildings.

In the spacing of facial features, we, on average, find 'pleasing'.

Architects have used it deliberately. Artists have played with it. Designers stumble into it unconsciously because it just 'looks right'.

And if you trade, you already know its cousins by heart: 38.2%, 50%, 61.8%, 161.8%. We call them Fibonacci retracements and extensions. If you've watched charts for long enough, you've seen prices respect these levels often enough that it becomes very hard to dismiss them as pure coincidence.

So from a simple pattern of numbers, born from repeated addition, we get spirals in nature, pleasing proportions in art and architecture, and statistically relevant levels in markets.

Geometry worth paying attention to, wouldn't you say?

## The Flower of Life and the Sound of Shapes

Now, shift to something that looks softer on the surface: the Flower of Life.

You've almost certainly seen it, even if you didn't know its name. It appears in temples, churches, synagogues, mosques, and in modern designs. At first glance, it looks like a cluster of evenly spaced circles overlapping to create a flower-like pattern of petals.

Constructing it is simple in principle:

Draw a circle. Using the same radius, draw another circle so that its centre lies on the edge of the first. Repeat this process, placing each new circle on the intersections of the others. Before long, you have a honeycomb of circles. Within that web, you can pick out familiar forms: the Seed of Life, the Tree of Life grid, hexagons, six-pointed stars, the Vesica Piscis (two overlapping circles forming a 'fish' shape), which appears all over Christian iconography.

The Flower of Life has been carved and painted into temples in Egypt, into medieval churches, into synagogues and mosques. Leonardo da Vinci studied its geometry. It turns up in cultures that, on the timeline we're given, had little direct contact.

Now switch fields completely and look at cymatics—the study of how vibration shapes matter.

Imagine a metal plate with a thin layer of fine sand on it, connected to a speaker. As you play different tones through the plate, the sand trembles and shifts. At certain frequencies, it suddenly leaps into beautifully ordered patterns: simple lines and divisions at low notes, more intricate mandala-like designs as the pitch rises. Do the same with water in a dish, and you see ripples and standing waves forming structures that, in some cases, look eerily similar to the diagrams found in sacred geometry.

In other words: sound—vibration—creates form.

Now hold a cymatic image next to a Flower of Life diagram. In some pairings, the resemblance is hard to ignore.

You could argue that ancient builders and artists intuited these patterns just by observing nature and sound, and then codified their understanding using geometry. Or, if you lean more mystical, you might say they were deliberately encoding their knowledge that vibration underlies reality.

Either way, a bridge appears between vibration, shape, symbol and state.

Suddenly, it makes sense that religious buildings—spaces whose whole purpose is to shift consciousness—are covered in patterns that echo cymatic shapes, Flower-of-Life-type constructions, and repeating ratios.

The vibrations of chanting, singing, organs, drums and bells move through spaces specifically designed to amplify and shape them. The geometry of the building and the geometry of sound interact. Your body stands in the middle of that interaction and responds.

You don't need to know a word of theory for that to happen. You just feel it.

## Compasses, Squares and Hidden Lessons

Now, invite the Freemasons into the room.

Say the word 'Freemason' and some people see old men in aprons and funny handshakes. Others start thinking about secret power structures and shadow governments. As usual, the truth is probably less dramatic than the rumours and more interesting than the PR.

The Freemasons are a fraternal organisation that emphasises community service, mutual support among members, and the moral development of individuals.

Whatever you believe about their influence, one thing is obvious if you look at their teaching tools: Freemasonry is obsessed with geometry.

Their most famous emblem—the square and compasses—sits on countless lodge walls, rings and buildings. In the centre is the letter G, interpreted as God, Geometry or both. Their tracing boards, used for instruction, are full of pillars, checkered floors, compasses, squares, staircases, stars, suns and moons.

The language of their rituals and stories revolves around architecture. Rough stones being smoothed. Temples being built not just from rock but from virtue and knowledge. Light as a symbol of understanding. Darkness as ignorance.

I'm not here to decode Masonic ritual. What interests me is the pattern: a group that explicitly uses the symbols of building and geometry as metaphors for personal and spiritual growth.

Then layer that on everything we've already discussed:

Pyramids aligned with stars.

Cathedrals built with harmonic ratios.

Temples and mosques tiled in endless geometric designs.

Mandala-like patterns appear in vibrating water and sand.

Fibonacci spirals in shells, flowers, galaxies—and price action.

Neural pathways are rewired under the influence of repeated thoughts and feelings.

At some point, it becomes harder and harder to maintain the old 'all just a coincidence' story with a straight face.

## Knowledge Kept Close

Now bring the Vatican back into frame.

Behind the public museums and guided tours sits the Vatican Apostolic Archive: rooms and vaults filled with manuscripts, letters, maps and artefacts gathered and guarded over centuries. Access is limited. Speculation about what's inside ranges from the mundane (political correspondence, historical records) to the dramatic (lost gospels, remnants of burned libraries, evidence of earlier civilisations).

You don't have to join every dot to notice a broader trend:

Powerful institutions, whether religious, corporate or governmental, tend to collect and guard knowledge.

You see it in imperial capitals, in secret societies, in corporate research labs, in intelligence agencies.

Combine that with what we see carved into stone and coded into symbols, and a simple question surfaces:

What if some of the deep pattern knowledge—the knowledge that connects geometry, sound, consciousness and matter—has always been held, to some degree, by a relatively small number of people, and simply never prioritised for mass education?

Not necessarily because there's a cackling villain in a volcano base, but because some knowledge multiplies power, and power tends to protect what multiplies it.

In trading, we're used to this idea. Institutions pay for data feeds and orderflow tools that retail traders never see. Quants build opaque algorithms that give their firms a structural edge. On the symbolic side, you could imagine a similar dynamic: complex teachings reserved for initiates, simplified stories for the crowd.

In this book, in a much smaller way, we're doing the opposite:

Taking ideas that are often wrapped in obscure language—sacred geometry, manifestation, vibrational states—and mapping them back to things you can see and use: charts, habits, posture, breath, belief, action.

## A Lens, Not a Cheat Sheet

At this stage, you might be half-waiting for a reveal: a single diagram that 'explains it all'—how to get rich, hack the universe, bend reality with your mind.

That's not how this works.

Sacred geometry is not a cheat sheet you memorise and then cash in.

It's more like a lens.

Once you start looking through it, you see connections differently. You glance at a spiral galaxy, a pine cone, a storm on a weather map, a fifteenth-century church window, the layout of three pyramids on a desert plateau, a cluster of candles on a price chart—and instead of filing them in separate mental boxes, you ask, "What patterns do these share? What are they expressing?"

At a practical level, that lens changes how you view your own life.

You stop reading your days as a bag of random events. You begin to suspect that the same forces that organise spirals, ratios and patterns 'out there' might also be relevant 'in here':

In how your thoughts spiral.

In how your habits repeat.

In how your emotions rise and fall.

In how your wealth grows or contracts.

You become more curious about structure. How is your day structured? How is your business plan structured? How are your relationships structured? How is your environment structured?

You notice that cluttered, chaotic spaces tend to promote cluttered, chaotic minds, while clear, ordered environments often support clearer, more ordered thinking. You notice that certain proportions simply feel better over time: too much work and no rest, and the ratio breaks; too much consumption and not enough creation and the pattern distorts; too much information and not enough integration and the system overloads.

Geometry starts to show up inside you as rhythms of effort and rest, input and output, solitude and community, risk and safety.

Seen that way, sacred geometry is not about drawing perfect diagrams in a notebook. It's about honouring balance and proportion in the way you build your life.

## The Stance, Not the Answer

So where does this leave us?

We've wandered through deserts and cathedrals, data feeds and diagrams. We've watched ancient builders align stone with stars. We've seen how sound sculpts matter into shapes that echo religious art. We've glanced at the symbolic language of Freemasons and at institutions that hide as much as they reveal.

We've seen Fibonacci and the golden ratio link shells, flowers, galaxies and market swings. We've nodded at the Flower of Life and watched cymatic patterns rhyme with its petals. We've felt how walking into a well-proportioned space shifts our state without a single word being spoken.

You might still be waiting for a neat definition: 'So what does sacred geometry mean?'

For me, it's less a sentence and more a stance.

Sacred geometry suggests we live in a patterned universe.

Not a rigid machine where everything is pre-written, and not a meaningless chaos either. A responsive system with preferences: for balance, for harmony, for certain proportions and rhythms.

Your mind, your body, your wealth, your relationships, your spiritual life—they all unfold within that patterned field.

You don't need to decode every symbol in a Masonic lodge, every carving on a temple wall, or every painting in the Vatican to benefit. You don't need to become a mathematician or a mystic.

You just need to accept that structure is not the enemy of freedom; it's the vessel for it. Proportion is not a constraint; it's what keeps systems from tearing themselves apart. Geometry is not abstract; it's the unconscious script your life is already following.

Once you see that, you can stop ignoring the shapes and start working with them.

You can design your days like a well-proportioned room instead of a chaotic storeroom. You can shape your habits like spirals that build upwards rather than loops that trap you in circles. You can approach charts as human geometry instead of random jagged lines. You can treat your own growth as a building project: foundations first, then pillars, then arches and openings.

The ancients encoded that idea in stone, symbol and song. Secret societies wrapped it in ritual. Priests wrapped it in scripture. Physicists wrap parts of it in equations. Traders, whether we admit it or not, sketch some of its outlines with our lines, boxes and ratios on charts.

You don't have to understand every expression.

You only have to understand this:

Life is not flat.

It is layered and patterned, and whether we call those patterns 'sacred' or not, they run deeper than we were encouraged to see.

Sacred geometry is one way that depth becomes visible.

From this point on, every time you see a spiral, a star, an arch, a mandala, or even a Fibonacci retracement that price respects to the tick, you have a choice.

You can shrug and call it luck.

Or you can nod, however quietly, and recognise that you've just glimpsed the fingerprint of something older, larger and more intelligent than any single human life, any single religion, any single system.

You can let that make you superstitious, or you can let it make you more respectful—of the patterns you move through, and of the patterns you are creating, every single day.

# Chapter 17: Different Meditation

**"Empty your mind, be formless, shapeless – like water."**

Bruce Lee

Doing nothing is the problem. At least, that's what it feels like to me.

You sit on the sofa, try to knock off, but your brain is a browser with a hundred tabs open. Charts you could be checking. Ideas you could be building. Conversations you're replaying. Things you should have said. Things you wish you hadn't. Your body is in one room; your mind is halfway across the planet, halfway through ten different futures.

For years, when people said "meditate," I heard "sit still and try not to think." The classic "count your breaths to ten and back" routine. It helped, but only up to a point. It felt like white-knuckled stillness—me trying to hold back a river with my bare hands, then judging myself every time a thought slipped through. That isn't focus; that's fighting yourself.

Somewhere in all this, I came across two Zen ideas that finally made sense to me: Muso no Ichi and Ichigyo-Zammai.

- Muso no Ichi, translates to 'the one without a peer' in basic terms, it encourages the practitioner to become the 'one' with their action, often leading to a heightened state of awareness or flow.

Ichigyo-Zammai represents 'one-action samadhi.' Immersing yourself completely in a single task or action, allowing distractions to dissolve, leading to an intense, deep meditative state, and concentrating on that one action.

Think of a swordsman. He doesn't narrate, "I will now block this strike." By the time that sentence is finished in his head, he's on the floor. He just blocks. Or he doesn't. There is no gap between seeing, deciding, and moving. That's Muso no Ichi—the experience of flow where the line between 'me' and 'what's happening' starts to dissolve.

I recognised it immediately, from my trading.

Those days when I'd done the prep properly. Levels marked. Risk calculated. No revenge-trading soundtrack running in the background. A setup appears. My hand moves. No inner debate. No begging the chart to confirm my emotional needs. Just: there it is. Click. Done.

Hours later, I'd look back and think, "That didn't feel like me trying. That felt like the system working through me."

That is Muso no Ichi. One thing is happening, clean.

The problem was that those days felt like accidents. Gifts. I had no way of calling that state in on purpose. My question became: how do I choose to be there more often?

That's where Ichigyo-Zammai came in.

In simple terms: you choose one thing, and you give it your entire attention. Not in theory. In practice.

One breath, one posture, one keystroke, one email, one candle or one chart. You get the point.

No multitasking. No half-checking your phone. No background argument replaying from yesterday. Just total engagement with a single action until, for a moment, it fills the whole horizon of your awareness.

If Muso no Ichi is the actual flow state, Ichigyo-Zammai is the training ground for it. You don't wait for 'the big fight' to discover focus. You practise on small, boring things until focus is a habit your nervous system understands.

That's when the penny dropped for me.

Most of what I'd been calling 'meditation' was really me trying to suppress experience. Sit here. Don't think. Fight every itch. Glance at the clock. Fail. Repeat. These Zen ideas weren't saying "don't think." They were saying:

"Pick one thing—and keep returning to it, gently, until, for a while, there is only that."

Muso no Ichi: the state itself.

Ichigyo-Zammai: the habit of returning.

Once I saw it like that, I realised I'd already been talking about the same engine throughout this book:

- Neville Goddard's 'living in the end' is just Ichigyo-Zammai aimed at a future state: stay with the feeling of the wish fulfilled until everything else becomes background noise.

- The observer effect is physics' way of saying: focus collapses probabilities. Observation selects a reality.

- When we spoke about 'frequency,' it was the same principle in different clothes: hold one state long enough and your system has to reorganise around it.

The logic is simple: you're training a muscle.

The same muscle that lets you stare at a chart for an hour without flinching. The same muscle that lets you sit with discomfort instead of lunging for cheap dopamine. And the same muscle that keeps you from spiralling in a losing streak or getting reckless in a winning one.

You're teaching your brain that it is possible to be here, with this, without splitting itself into ten directions.

Here's an example of the opposite.

Imagine an engineer was staring at a design he'd drafted a hundred times before. The calculations checked out. The tolerances were fine. The safety factors were conservative. On paper, it was the kind of drawing you could sign off in your sleep.

And yet, he freezes.

His mind sprinted to the last project where something went wrong on site, the email from a client months ago picking holes in his work, an imaginary future where this one decision somehow collapsed a bridge and ruined his career. He wasn't looking at steel and load paths anymore; he was looking at ten years of identity hanging on a signature.

That's the opposite of Muso no Ichi.

That's 'I' over here and 'action' over there, with a thick story in between. So thick you can't see the present moment through it.

You need to be able to ask

"What is actually here, now, without me in the story?"

Not the past or not the future. What is the now?

What caused the doubt?

People tend to manufacture their own condition for inaction. Blocking themselves from achieving whatever it was they set out to do.

That's why Ichigyo-Zammai matters: it's the practice of bringing your attention back to one thing until, sometimes, it just stays.

Neuroscience has its own label for it: Flow state.

Athletes, coders, musicians, and surgeons talk about it all the time:

- Time distorts.

- Self-consciousness fades.

- Action and awareness blend.

- The quality of what you're doing goes up precisely because you're not trying so hard.

Mihaly Csikszentmihalyi mapped the conditions:

- Clear goal.

- Immediate feedback.

- Challenge just beyond comfort, but not so far you're panicking.

- And—this is the important part—removal of distraction.

In other words: Ichigyo-Zammai. Pick one thing. Let it matter. Stay with it.

Modern 'deep work' routines and NSDR (non-sleep deep rest) are just updated names for it:

- Phone in another room.

- No switching windows.

- Lie down, let the body drift towards sleep while keeping a thin thread of awareness on the breath.

Different clothes. Same principle. You're nudging your brain out of jittery, fragmented beta and into a steadier, more coherent pattern where Muso no Ichi can actually show up.

The breakthrough for me was realising that meditation doesn't have to mean sitting on a cushion trying not to think. It can be:

- Walking down the road and feeling every footstep.

- Drinking a cup of tea and actually tasting it—heat, flavour, weight of the mug in your hand.

- Writing one paragraph with full presence, instead of half-writing it while planning the next six chapters.

- Reviewing one trade properly, instead of skimming ten half-heartedly.

It can be turning the most mundane job into practice: brushing your teeth, folding clothes, cleaning the car. Each one a chance to say, "This. Just this. For thirty seconds. Then the next thing."

Over time, something quiet starts to shift.

Your nervous system screams a little less when there's nothing to distract it. The urge to check your phone every thirty seconds softens. You become slightly more willing to sit in silence and see what floats up without drowning it immediately in noise.

That willingness is where the real work starts.

Because once you can sit quietly without panicking, you can finally do what I talked about in the hypnosis chapter: ask better questions and hear the answers.

What am I actually afraid of right now?

What "I am" sentence is running this behaviour?

What would one small, brave action look like today?

At that point, 'meditation' stops being a break from life and becomes a tool for designing it.

Not chasing mystical fireworks. Not hunting for visions. Just using old techniques like Muso no Ichi and Ichigyo-Zammai—to train the simplest and the hardest skill:

Staying with one thing long enough for it to change you.

In practical terms, it might look like this:

You sit down, close your eyes, and decide:

'For the next ten minutes, my breath is my entire world.'

Thoughts come. You notice them and return. No drama. No self-abuse. Just the quiet insistence of someone resetting a compass over and over. That's Ichigyo-Zammai.

Another day, you're in the middle of work—writing or coding, making a plan for a coaching session. A moment comes where, for once, you are fully there. No background commentary. No third-party narrator in your head. You realise you haven't thought about yourself in a few minutes. You've just been... doing.

That's Muso no Ichi.

The more often you create the conditions for those moments, the more your brain starts to prefer them. It learns that depth is less exhausting than scatter. That quiet focus is actually more comfortable than constant stimulation. That, ironically, you feel more like 'yourself' when you stop narrating yourself and simply act.

For a trader, entrepreneur, or anyone trying to change their life, that's an edge.

Not just because you'll make fewer stupid mistakes—although you will.

Not just because your work will improve—although it does.

But because you finally have somewhere to stand inside yourself that isn't at the mercy of every notification, every mood swing, every external result.

You start to experience, in tiny, repeatable ways, what the Zen teachers were pointing at all along:

There is a place in you that can do one thing completely.

Breathe.

Act.

Observe.

Decide.

Touch that place even briefly—in a trade, in a conversation, in a choice about money—and you're no longer a pawn being kicked around by invisible hands.

You are, in that moment, the player.

And every time you sit down, close your eyes, and deliberately give one thing your full attention, you are strengthening the player's hand.

Don't treat it as an escape from the game.

Treat it as remembering who is actually holding the pieces.

# Chapter 18: Other forms of Psychology

**"Keep your mind fixed on what you want in life: not on what you don't want."**

Napoleon Hill

I entered the realms of tech and trading somewhat unexpectedly, and these two fields have significantly influenced my perspective. Technology taught me that everything is built on patterns, logic and code. Trading taught me that you can have the perfect logic and still fall apart because of what happens in your head when money is on the line.

Why do some people run the same 'code' and get completely different outcomes?

Some traders fail. Some businesses go bust.

On the surface, it's easy to say, 'Bad strategy,' or 'The product wasn't good enough.' But the more time I spent watching real people in real markets, the more I started to wonder whether that explanation was too shallow.

What if it isn't the strategy?

What if it isn't the product?

What if the real failure sits deeper—in a program running quietly in the subconscious?

I can tell you now: it is not easy to think like a professional trader.

Most people imagine it's all about finding the magic entry, the perfect indicator, the secret source of information. But over the years, in our Discord community, I watched something that made me question that completely.

I had given hundreds of students the same education. The same content. The same tools. The same time and access to me. They sat in the same live sessions, watched the same examples, and had the same rules written down.

Yet when payouts started to appear, the stories split.

On one side, I had students sending me screenshots of $20,000 months. Calm, confident messages: "Just passed another challenge," "Withdrew this month, moving to the next level," "Finally starting to compound."

On the other side, I had traders who were still losing trade after trade. Same charts. Same rules. Same mentor. Very different lives.

At first, I tried to fix it the way all traders do. We went back over the entry criteria. We fine-tuned risk management. We talked about position sizing, stop placement, and win rates. We re-checked the 'logic'.

But the pattern wouldn't go away. The ones who were winning kept winning. The ones who were stuck stayed stuck.

What separated them?

I wanted an answer for them, not just a theory for me. So I did what I always do when something doesn't add up: I went digging.

I started reading more about psychology, assuming there had to be some magic sentence or mental trick that would flip a switch. Something I could say to a struggling trader and—click—their brain would start working like a pro's.

But the more I read, the more I realised there was no single magic line. There were threads. Patterns. Deep ones.

It wasn't until I started recognising the topics we've been talking about in this book—mindset, faith, visualisation, self-image, subconscious programming—showing up again and again in psychological language, that things really began to make sense.

The problem the struggling traders had was not a small fix. It wasn't even a 'trading' problem in the narrow sense. It was a self-problem. A psychological problem.

And that pushed me deeper into psychology than I'd ever gone.

If you type 'psychology' into Google, you get a wall of categories: clinical, counselling, cognitive, behavioural, developmental, and social. It looks like a maze. Techniques, therapies, acronyms, models.

Underneath all that complexity, one concept kept popping up, quietly but consistently.. Self-actualisation.

It wasn't a term I knew well at the start. It comes from Abraham Maslow—the same Maslow who gave us the 'hierarchy of needs' pyramid you might have seen in school. At the base: food, water, safety. In the middle: love, belonging, esteem. At the top: self-actualisation.

Maslow described self-actualisation as the realisation or fulfilment of one's talents and potentialities. In simpler language: becoming who you're capable of being.

It immediately took me back to Henry Ford: "Whether you think you can, or you think you can't—you're right."

Self-actualisation is Ford's quote, written out as a life project. If you think you can't, you'll never try, or you'll sabotage every attempt. If you think you can—and you back that up with action—you gradually evolve into someone who proves you right.

So when I looked at my two groups of traders through that lens, the picture sharpened.

The problem was simple on the surface: the individual. But behind that 'simple' label were layers of belief, habit and identity.

Think of the basic trading example we've used before: a strategy that wins 50% of the time but uses good risk management. You risk 1R to gain 2R. You only need to be right half the time to grow the account. Mathematically sound. You can write it out in ten seconds.

Now imagine Trader 'A' and Trader 'B' both use that system.

Trader 'A' takes a series of losses. Ten, fifteen, twenty in a row. It hurts. But somewhere inside, they still hold the belief, "I am a profitable trader working a profitable edge." They see the losing streak as part of the distribution, not a personal verdict. They adjust, review, maybe take a break, but they don't rewrite their identity.

Trader B takes the same series of losses. They feel every one of them like another nail in the coffin of their self-worth. After trade 20, when they say, "I've lost twenty trades in a row," there's a sentence hiding behind it:

"Trade 21 will be a loser too."

They don't say that out loud, but they feel it. Their mind is already leaning into failure. They have placed their energy, their faith and their belief into the expectation that the next trade will be the same story.

When that's the case, what are they looking for on the chart, subconsciously?

They're looking for what fits their story: another losing trade.

They rush entries. They ignore their rules. They skip higher-probability setups because a part of them is whispering, "What's the point? You lose anyway." Or—and this sounds strange but it's very real—they take reckless, oversized trades that all but guarantee another hit, because their nervous system has become more familiar with the pain of losing than the discomfort of winning.

This is where behavioural psychology starts to make sense of it.

On paper, behavioural psychology reduces human action to reinforcement and habit. You do what you're rewarded for. You avoid what hurts. Most of the time, that's true.

But after twenty straight losses, 'losing' has become weirdly familiar. The brain, which prefers predictable pain to unfamiliar possibility, starts to treat the idea of a win as dangerous.

It sounds insane—but at a subconscious level, a big win means change. Change means stepping into a new identity. New identity means risk: new expectations, new responsibilities, new people, new pressure.

So your brain quietly makes losing feel safer.

You can see, now, why quick fixes don't work.

You can also see why so many trading books that try to tackle 'psychology' by saying, "Stick to your plan," or, "Be disciplined," miss the depth of what's actually happening.

For me, this is where everything started to link up.

Positive psychology isn't just about wanting to win, or chanting optimistic slogans. It's about deeply believing that the win is possible, and then repeatedly entering the emotional state of that win as if it has already been achieved—something we've talked about with Neville Goddard and 'living in the end'.

Back in Chapter 1, when I talked about the wealth mindset, I said that wanting to be wealthy isn't enough. You have to shift the mind, the emotions and the belief at a fundamental level. What I was really pointing at, even if I didn't use the word then, was self-actualisation.

The traders who were winning—and winning big—had a different aura about them. They knew they could lose the next trade. They weren't naive. But they didn't carry those potential losses as a personal indictment. Their confidence and momentum meant they were more likely to wait for better setups, size sensibly, and bounce back quickly. Their internal self-image was, "I am a trader who wins over time," not, "I am a loser waiting for more proof."

In other words, they had begun acting from a higher level of self. They were becoming the version of themselves who could handle success.

In this chapter, I want to nod towards some other branches of psychology—not to drown you in jargon, but to show you how they line up with what we've already explored.

Take cognitive psychology and its practical cousin, cognitive-behavioural therapy (CBT).

At its simplest, CBT says this: your thoughts influence your emotions; your emotions influence your behaviour; your behaviour feeds back into your beliefs. If you change the way you think about events, you change the way you feel and act.

Translated into trading: if you see a loss as proof that "I'm useless and can't do this," you feel hopeless and either quit or gamble. If you see the same loss as 'expensive information'—feedback about your plan—you feel curious or determined and go back to study.

Same external event. Different internal story. Different outcome.

If you've read any CBT material, you'll recognise some of the thought patterns they list as problematic: catastrophising ("This one loss means I'll never be profitable"), black-and-white thinking ("I'm either a genius or an idiot"), personalisation ("The market is out to get me"), fortune-telling ("I just know this will be a loser").

Every single one of those shows up in trading chats every day.

Then there's schema therapy, which digs one layer deeper. It talks about 'schemas'—core themes and patterns we develop early in life about ourselves and the world.

"I am unlovable."

"I am a failure."

"People always leave."

"The world is dangerous."

Those schemas become lenses. Once in place, we tend to interpret every event in a way that confirms them.

A trader who deep-down believes, "I am a failure," will find ways to bend their results back to that narrative. Blow-ups, missed opportunities, and sabotaged systems. Not because they like pain, but because their system is trying to keep its internal map consistent. The same for founders and business owners – "it's hard for me" is the program most deployed.

Then you have approaches like Acceptance and Commitment Therapy (ACT), which say, in essence: stop trying to get rid of every uncomfortable thought and feeling, and instead learn to notice them, accept their presence, and still act according to your chosen values.

In our context, that might look like a trader feeling fear, doubt, or the urge to overtrade—and instead of believing those feelings, they say, "There's fear again," breathe, and still follow their plan. They stop waiting to "feel ready" and start acting in line with the identity they're building.

All of these forms of psychology circle the same truth from different angles:

If you don't work on yourself, no system will save you.

If you do work on yourself, even an average system can become a vehicle for serious growth.

Now, zoom out.

Imagine anyone—me, you, a stranger—walking into a psychiatrist's office. The psychiatrist's official role is to sit, listen, ask questions, and try to help you untangle what's going on inside your head. Many of them are very skilled at that.

But there's a catch.

They're a stranger.

And unless you're drunk, high, or already very self-aware, your guard goes up around strangers. You edit. You tell versions of your story that make you look a certain way. You hide the parts you're ashamed of, or that don't fit your own narrative.

That doesn't make therapy useless—not at all. It just highlights something important:

There is a limit to what someone else can do for your mind if you are still hiding from yourself.

What if, instead, you could train yourself to be fully aware and honest about your state, to see your own patterns as clearly as a good therapist would—and then, crucially, drop your guard with yourself?

What if you could learn to notice your thoughts without instantly believing them, recognise your old schemas as stories rather than facts, and gently challenge your own "I am..." statements the way a psychologist would?

What would that kind of self-honesty and self-awareness give you?

For a start, it would give you the ability to bring your subconscious patterns into the light on a daily basis, not just in a 50-minute appointment once a week.

It would let you see, in real time, when you start setting up another self-fulfilling prophecy. "I've lost twenty trades; therefore, I will lose the twenty-first." "Every time I get ahead, something goes wrong." "People like me don't..." and so on.

It would allow you to pause there and ask: "Is this thought an absolute law, or is it a script I've rehearsed? What would I do right now if I didn't believe this sentence?"

You would begin to act less like the old program and more like the person you actually want to become.

Psychology in all its branches—clinical, cognitive, behavioural, positive, existential—ultimately points toward the same task:

Know yourself, so you can stop being run by parts of you that never got updated.

In the context of this book, that looks like:

Noticing when your wealth mindset collapses back into scarcity thinking.

Catching your faith slipping into external blame instead of internal responsibility.

Seeing when your visualisations are half-hearted because you secretly think "not for me."

Recognising when your body is stuck in fight-or-flight, and your decisions are reflecting that.

Becoming aware of the hypnotic scripts you've inherited and repeated: "I can't," "I don't," "I never."

Then, using the tools we've talked about—"I am" statements, visualisation, breath, movement, environment, self-hypnosis—you begin to re-write.

That's psychology, too.

It's just psychology you are applying to yourself, in your own language, in your own life.

When I look back now at those two groups of traders—the ones who took off and the ones who stayed grounded—it's obvious that the dividing line wasn't intelligence or information.

It was a willingness.

Willingness to look in the mirror.

Willingness to question old identities.

Willingness to feel discomfort while building a new self.

The winning traders, whether they could name it or not, had started walking a self-actualisation path. They were not just learning entries and exits; they were becoming people who could handle success without running from it.

The struggling traders, however talented, were often still fighting yesterday's battles: trying to prove someone wrong, trying not to repeat a parent's mistakes, trying to avoid shame. The charts were just the screen on which those patterns played out on.

In the next chapter, I'll pull all of this together—how wealth mindset, faith, visualisation, science, ancient structures, frequencies, bodywork, hypnosis, and these 'other forms of psychology' all condensed into one picture in my mind.

But for now, if you take nothing else from this chapter, take this:

You can go to therapists, coaches, gurus and mentors—and in the right context, they can help enormously. But the deepest psychological work is ultimately about working on yourself, with yourself.

Training yourself to be aware.

Dropping your guard with yourself.

Learning to see your own subconscious patterns.

Then choosing to re-align them with who you are becoming, not who you used to be.

In a world where almost everyone is trying to change their results without changing their inner code, that is an extraordinary power to claim.

You know where else psychology can be affected, even manipulated?

MUSIC...

A distinct pattern emerges when you examine individuals who perceive reality in unique ways. The information doesn't trickle in gradually or come as fragmented thoughts that need to be forced into coherence. Instead, it arrives fully formed and well-structured, as if someone had meticulously arranged the entire package before their arrival.

The classical composers knew these principles, whether they would have put it in those words or not.

Bach built fugues with pure mathematics. Voices interlocking at exact intervals, themes returning at precise points, structures that mirror Fibonacci proportions and recursive patterns. Listen to a good performance of a Bach fugue with your eyes closed and you can almost feel your brain reorganising itself—because it is. Order and complexity, woven together. If you ask Alexa for this you will need to ask for 'Bach Fugue in G minor, BWV 578'.

He never explained how he knew. He just did it.

Mozart was more direct about his process. He said the music arrived complete. His symphonies and concertos came to him already orchestrated, already finished. His job was simply to write down what was being transmitted. Not trial and error, not 'let's see what chord fits here.' Reception, just like Tesla and his inventions.

Beethoven may be the most bizarre example. He went deaf. Not 'a bit hard of hearing.' Completely deaf in the last stretch of his life. And it was during that period that he produced some of the most complex, emotionally devastating works in the world.

The Ninth Symphony was written when he could hear absolutely nothing of the outside world.

Go listen to it!

How did he do this? He spoke of hearing from within. A level beyond physical sound. The music existed first as a pure form—structure, proportion, movement—before it ever turned into vibration in the air.

A man in total silence, 0 external input, and he produces a symphony that still makes crowds cry two hundred years later. That is access to something most of us don't even know exists, never mind learn to use. A field? A deeper layer of the mind? Call it what you like. It's there.

Wagner took that same understanding and weaponised it.

He wasn't trying to entertain. He was trying to transform. He wrote operas that lasted four or five hours. Harmonic progressions that never quite resolved, keeping the psyche suspended on a hook. Musical themes that returned and returned, attached to characters and ideas—what we now call leitmotifs—building layers of association in the unconscious.

Nothing was accidental. Everything was calculated. It worked. Nineteenth-century Germany did not walk out of those opera houses the same. An entire nation had its emotional vocabulary shifted by sequences of sound. Wagner showed what's possible when you understand vibration for real, not as a New Age slogan, but as a technology.

Then came the film industry.

John Williams, Hans Zimmer, Ennio Morricone—these guys are not 'just' composers. They are emotional engineers. They studied all of this. They know exactly what they are doing with every interval, every swell, every silence.

There are harmonic shapes that trigger fear before you can explain why. Your body reacts; your mind catches up later. Others create instant nostalgia—a certain chord change, and you are eight years old in your grandparents' house again. Others generate tension you feel in your stomach and shoulders, even if the scene on screen is just two people talking.

Then you have the blunt instruments that became legend. The music from Jaws. 2 notes, just 2, alternating in semitones. And an entire generation developed a fear of entering the ocean. Children who have never seen a shark in their lives tremble when they hear that sound. 2 notes, programmed terror. This is power over the human mind. real power, measurable, replicable.

Once you see that, you can't unsee it.

You watch Star Wars and hear John Williams attach a specific theme to the Force. Over time, the moment that melody plays, you feel hope, destiny, something larger moving—even if it's just over a shot of empty space. You watch Inception and feel the BRAAAM of Zimmer's foghorn-like blasts, and your nervous system goes into 'something massive is looming' mode. You hear the first bars of your national anthem and your posture changes automatically.

None of that is random.

What does this have to do with money, wealth or mindset?

Everything.

Because it proves, in a way nobody really argues with, that:

1. We are programmable at a deep level by rhythm, pattern and repetition.

2. The programming often bypasses our logical mind and lands straight in the body.

3. Once an association is built (two notes = shark; certain chords = noble sacrifice; certain jingles = "I want that product"), it runs on autopilot.

You and I might never write a symphony. But we are doing something very similar to ourselves all day long.

You repeat:

"I'm unlucky."

"I always mess it up when it matters."

"People like us don't get rich."

"If I have more, someone else has less."

"Rich = cold and selfish; poor = warm and good."

You do it with words. You do it with the 'soundtrack' you play in your head when you look at your bank balance or your open trades. You do it when you tighten your jaw every time a bill arrives, pairing money with tension.

Over the years, those phrases, tones and postures become your private soundtrack. Your own leitmotifs.

Now, not just vibrationally, but in other forms hidden in plain sight.

Wealth moves into view? Your inner orchestra starts playing the theme called 'danger, don't get your hopes up.'

New opportunity arises? The 'I always blow it' motif sneaks in under the scene.

Time to raise your prices? Up comes the 'rich people are evil/greedy/shallow' theme from Titanic and Avatar and every other story where the wealthy are the villains.

You don't think those movies affected you. But they did. They rode on top of music engineered to embed stories in your nervous system.

And because the emotion was real—your heart really pounded when the ship was sinking, you really felt indignation when the miners in Avatar destroyed the forest—your subconscious tagged those stories as 'important.' It saved the associations.

Rich = Bad.

Powerful = Cruel.

Ordinary = Noble.

Those with the ship = Cowards.

Those in the water = Heroes.

So, when you start doing work to become wealthy, your conscious mind is reading Neville Goddard, Kiyosaki and this book. But your subconscious still hums the old themes. And when push comes to shove, the soundtrack wins.

The point here is not to make you paranoid or to tell you to stop watching films.

It's to show you that vibration—sound, rhythm, pattern—is a real, tested, weaponised way of moving human emotion and expectation. Governments know it. Propagandists know it. Advertising agencies know it. Film studios know it.

The only person who tends to pretend it's 'just entertainment' is the individual viewer.

If John Williams can make you feel awe for a fictional energy field with four notes of a French horn... you can absolutely make yourself feel calm and capable around money with a handful of chosen sentences, said in a particular rhythm, in a particular state, over and over.

If two notes can make a child who has never seen a shark fear the sea... your daily muttered "I'm terrible with money" can absolutely make an adult self-sabotage every promising opportunity.

The difference is intention, and composers use these tools to serve a story someone else is paying them to tell.

Even if you dig a little deeper, the same system is at play with the visuals and story. Both Avatar and Titanic made billions because they tapped into a pre-existing emotional pattern in the collective: a deep suspicion that wealth and power are inherently corrupt, and that 'people like us' (middle class, working class) are the victims of their greed. That story lands because, on some level, it feels true to many people's lived experience of inequality and corporate abuse.

The problem, psychologically, is that when this narrative is absorbed uncritically, it becomes part of your inner code. If 'rich' is consistently equated with 'cold, cruel, exploitative,' and 'good person' is consistently equated with 'broke, struggling, or sacrificial,' then your subconscious quietly links your own desire for wealth with becoming the villain. Middle-class viewers leave entertained but also reinforced in a very old schema: "If I stay like this, I'm on the side of the good; if I really break through, I must be selling out." That's a disaster for a wealth mindset. Your conscious mind might be reading books about abundance, but your emotional body has spent decades cheering every time the underdog beats the rich villain and every time the rich villain is punished. Guess which side your subconscious will try to keep you on.

Films and stories like these function as mass hypnosis scripts. They use powerful imagery,  and of course, the music, playing on emotions to anchor certain associations: money with coldness, industry with desecration, success with selfishness. Over time, that becomes vibrational conditioning. You don't just 'think' rich people are bad; you feel a subtle resistance, disgust, or guilt around the idea of being truly wealthy yourself.

That feeling is often what sabotages you right at the threshold—undervaluing your work, over-giving, pulling back when opportunities appear—because on a deep level, you don't want to step into the role you've been trained to hate. The antidote isn't to pretend exploitation doesn't exist; it's to consciously separate wealth from character in your own mind, and to build a personal model of 'rich and deeply ethical,' 'powerful and profoundly kind,' so your subconscious has new roles to aim at instead of only the movie villains.

On top of all that, we now live inside the most sophisticated propaganda machine ever built: targeted data. Multinational companies and governments don't just blast one soundtrack at everyone; they farm your clicks, pauses, searches and scrolls to compose a personal score for you. Google, Meta, TikTok and the rest learn exactly what images, phrases and offers keep you engaged. They know when to show you the low-interest (on the surface) credit card, the 'buy now, pay later' button, the luxury item you've half-fantasised about but don't actually need. Your feed becomes a hypnosis script tuned to your specific fears and cravings. If your inner code screams scarcity, you'll see endless adverts for quick fixes and debt; flip that code—truly start to embody "I am someone who builds wealth, I am capable of owning X" —and the very same system starts serving you different images: your dream car, the house you'd actually love, tools and opportunities that match a higher self-image. The external machine doesn't change; your signal does.

And as your signal shifts, what the machine reflects back at you shifts with it.

# Chapter 19: Connecting the Dots

***"The mind is everything. What you think you become."***

**Buddah**

When I first left school at fifteen, I didn't imagine I would one day be sitting in an old French farmhouse, surrounded by piles of books about business, psychology, religion, geometry, ancient temples, and declassified CIA documents, trying to stitch them together into a single story.

Back then, life was simple in a very particular way. You learned a trade, you went to work, you got paid, you tried to get by. The idea that your thoughts could shape your reality, that beliefs could bend the trajectory of your life, that numbers and symbols might be hiding in plain sight all around you—that would've sounded like fortune-cookie nonsense.

But life has a way of dragging you into questions you never planned to ask.

When I look back over the chapters we've walked through together, it doesn't feel random. It feels like what happens when you're staring at a messy chart for days, convinced it's just noise, and then suddenly the structure appears. The pattern that was there all along.

This chapter is about that pattern.

It starts small. A trade that goes right. A trade that goes wrong. Two students who get the same lesson and walk away with completely different results. An odd dream that turns into a real invention. A pyramid that refuses to behave like a simple pile of stone. A PDF on a government website talking, in all seriousness, about out-of-body experiences.

You start wondering if maybe—just maybe—the surface story of life is not the whole story.

This book is the result of that wondering.

Now, at the end, I want to tell you the story again, from a slightly higher vantage point, so you can see the pattern that's been forming underneath everything.

It begins, as many stories do, with money.

I've taught enough people now to stop pretending it's just about systems.

When I started taking on students, I did what any honest educator would do: I gave them what I knew worked. Structure. Risk–Reward. Entries. Exits. Case studies. Examples. I poured hours into Discord sessions, live trading, Q&As, and breaking down the same principles in different ways so that anyone, if they applied them, could start building consistency. The same thing with consulting, what worked, what didn't, understanding the why and evolving the systems and processes.

The first surprise was how quickly some of them took off.

"Lewis, I've just got an investment for my company."

"Lewis, I can't believe it—I've closed out a superb year."

These were not super-geniuses. They were ordinary people prepared to learn and act.

The second surprise—and the one that bothered me more—was how others, who had been in the same sessions, with the same PDFs and the same logic, were still blowing accounts, not gaining more sales, just stuck in perpetual no-man's land.

Same teaching.

Different outcome.

If this had happened once or twice, I could have blamed luck. But after you watch the same pattern play out with hundreds of people, you start asking sharper questions.

What if the main problem isn't strategy at all?

What if the main problem is personal beliefs?

That's where Chapter 1 came from.

We go back to Henry Ford's line: "Whether you think you can, or you think you can't—you're right." It sounds like something on a gym wall you glance at and forget. But when you lay it next to actual bank statements and trade journals, it stops being cute and starts looking like a law of human behaviour.

The student who thinks, deep down, "I'm not the sort of person who wins," finds a way to prove that true, even with a profitable system in their hands. The student who gradually comes to believe, "I can handle this; I can learn this; it's just a matter of time and practice," behaves differently long before their equity curve reflects it.

That is where the wealth mindset is born.

It is not a mood. It is not a slogan. It is a fundamental shift in identity: from 'I hope' to 'I am becoming.' And eventually, to "I am."

Once you see that belief is the engine under the bonnet, it becomes impossible to keep the rest of life in neat little compartments.

You can't say that belief matters in trading, but not in health, or relationships, or business, or anything else that requires you to show up consistently when you don't feel like it.

In Chapter 2, we went after belief itself.

We started with a line from Exodus: "I AM WHO I AM." On the surface, it's a statement from a god to a prophet. But read through the lens of psychology, it becomes a template. Every "I am..." is a command to the subconscious. I am broke. I am bad with money. I am disciplined. I am a learner. You don't just describe yourself with those phrases. You instruct yourself.

That's where faith and fortune begin to converge.

Not faith as in, 'Sit still and hope a higher power fixes everything,' but faith as in, 'I trust that if I move, things will move with me. I trust that if I walk, the path will appear under my feet.' A faith that sees God, or the universe, not only as something 'out there' but as something moving through you.

Combine that with personal responsibility, and you get the mindset we talked about: 'I am supported—but I am also responsible.'

The moment you stop outsourcing everything to fate, government, markets, or gurus, and start claiming your own "I AM", the game changes.

In Chapter 3, I added the next layer: visualisation.

Visualisation, used properly, is not about staring at a mood board and getting a small buzz before you sigh and go back to 'reality.' It's about using the way your brain actually works. fMRI scans show that vividly imagining an action lights up many of the same regions as performing it. Athletes know this. Musicians know this. Good performers of any kind know this.

Neville Goddard called it 'living in the end.' He argued that if you want a different future, you must first inhabit it in imagination. Not as wishful thinking, but as a felt state. Smell the sea air on the yacht. Feel the calm of closing the trade by plan. Feel the pride of paying off the debt. Not someday. Now.

When you do that repeatedly, your nervous system starts to treat that end state as familiar. Then, when choices arise in the present, you have a reference point: 'Does this belong to the version of me I've been rehearsing—or to the old one?'

Visualisation without action is daydreaming. But action without any inner rehearsal is like stepping into a ring having never shadow-boxed a round in your head. We need both.

Chapter 4 anchored these ideas in science.

We looked at how something as simple as forcing a genuine smile can change your state and your chemistry—the facial feedback hypothesis. We talked about body language and how your posture and tone broadcast your internal state, shaping how others respond. We brought in NLP to show how internal language and imagery form your personal 'code' can be altered or adjusted.

Then we zoomed into quantum experiments like the double-slit, not to turn you into a physicist, but to give you a metaphor: before observation, possibilities are many; after observation, one is realised. Before a decision, your life contains several potential paths; after you commit energy and focus, something 'collapses' into experience.

Whether you see that as mystical or mechanical, it underlines the same principle: where you place sustained attention matters.

We also demystified 'vibration' and 'frequency.' Fear and scarcity literally produce different heart rhythms, hormonal cocktails and brainwave patterns than calm confidence. Those internal states change how you behave, which changes what you experience.

The big idea: attraction is not magic. It's momentum.

Thought → state → behaviour → outcome

repeated until the snowball runs on its own.

At that point, we could have stayed tidy: mindset, science, performance. But life isn't tidy, and neither is truth.

In Chapters 5 and 16, we walked around pyramids, cathedrals, stone circles and temples. We looked at the absurd logistics of the Great Pyramid, the eerie emptiness of its inner walls, and the precision of its alignment. We looked at megaliths in South America fitted tighter than modern masonry, at Baalbek's monster blocks, at Göbekli Tepe rewriting timelines, at Chinese pyramids buried under trees.

We asked uncomfortable questions: Why do religious buildings across cultures use the same kinds of ratios and forms? Why does sacred art across continents echo the same Flower-of-Life patterns and spirals? Why does vibration—sound played through plates of sand or dishes of water—create shapes that look like those same designs?

Fibonacci, the golden ratio, the Flower of Life, cymatics, Masonic compasses and squares, star alignments in stone... they all begin to hint that reality has preferences. That form is not random. That proportion matters.

I wasn't trying to turn you into an archaeologist. I was trying to widen your lens. To show you that the same instincts you use when you look at a chart—spotting rhythm, ratio, symmetry and imbalance—might be relevant when you look at your own life, and even the universe.

Chapters 6 and 10 showed how this isn't confined to stone.

We looked at frequency in the literal sense: brainwaves, sound frequencies like 432 Hz or 528 Hz used for mood and meditation, the way certain tones and rhythms move your nervous system. We explored crystals as structured matter that respond predictably to pressure and electricity, and how modern science is now using quartz as a long-term data storage medium—essentially doing, in labs, what shamans and mystics have poetically claimed for centuries: that crystals can 'hold information'.

We walked through Tai Chi: rooting, centre, song, intent. We watched how slow, precise, seemingly soft movements train the body to be grounded, relaxed but ready, and guided by a clear line of intent. We saw how push-hands mirrors the way markets move stiff, ego-driven traders around, and how a flexible, responsive trader behaves more like a good internal martial artist than a stubborn brawler.

We even glanced at Dim Mak and pressure points—not to glorify harm, but to illustrate a tough truth: if you know the structure of a system, you can influence it greatly with small, well-placed interventions. The same map can heal or hurt, depending on intent.

The message underneath all this was the same:

You are not just a mind floating around with 'mindset tricks.' You are an energy system in a physical structure in a patterned universe. How you stand, breathe, move, and focus is not incidental. It is the way you 'speak' to that universe.

Chapters 7 and 11 pulled technology and altered states into the picture.

We looked at circuit boards, IF–THEN logic, Mandelbrot fractals, Fibonacci in markets, AI and quantum computing. We asked whether mathematics might be the bridge language between what religious people call 'God' and what engineers call 'code.' We noticed that in both cases—scripture and software—the story starts with a word, a command, a line of information: 'In the beginning was the Word,' 'In the beginning was the code.'

Then we stepped into hypnosis and drugs. Not to glorify either, but to show how suggestible the mind is. We saw how advertising, media, education and even casual conversation can function like low-grade hypnosis, installing "I am" statements and expectations without you realising. We looked at how altered states—from meditation, breathwork, trance, or substances—open that door wider for better or worse.

The point was not to make you paranoid. It was to make you responsible.

If the mind can be programmed—from outside or inside—then the question is not, "Am I being influenced?" but, "By whom?" And, "Am I going to learn to influence myself on purpose?"

Chapters 8, 9 and 17 brought us back to the everyday grind: motivation, Stoicism, mind–body connection, and other forms of psychology.

We talked about what happens when the toast burns, the coffee goes cold, and the petrol light comes on. How easy it is to let minor annoyances drag your state down, undoing an hour of morning visualisation in seconds. Stoicism gave us a simple compass: control what you can (your response), accept what you can't (the burnt toast), and keep walking.

We explored how thoughts become chemistry, chemistry becomes posture, posture becomes behaviour, and behaviour becomes outcome. We saw that if you eat, sleep and move like someone who doesn't care about tomorrow, your brain can't support the beliefs and focus of someone building a serious future.

We touched on self-actualisation: Maslow's idea that at the top of human development sits the drive to become fully yourself—to realise your potential. When I looked at winning traders and builders through that lens, they all had one thing in common: they were willing to grow into a bigger version of themselves. The losing traders often weren't fighting the market; they were unconsciously defending an old identity.

Chapters 12, 13 and 14 together show that by unlearning hypnotic "I am" scripts, using communities and mentors as scaffolding rather than crutches, and recognising you're operating in a deeper, patterned reality, you can deliberately redesign your identity—the inner belief system that either opens up or puts a ceiling on your capacity for wealth and meaning.

The more psychology I read—CBT, schema therapy, ACT, positive psychology—the more it confirmed what this book keeps saying in its own language:

You have scripts running. They can be changed.

You have patterns playing out. They can be redesigned.

You have an inner core that is more than your conditioning. It can be listened to.

So where does that leave you, at the end of all this?

You've walked with me through wealth mindset, faith and "I AM", visualisation, science, geometry, pyramids, frequencies, crystals, Tai Chi, mudras, Dim Mak, CIA documents, hypnosis, Stoicism, sleep and food, self-actualisation and self-sabotage.

You could put the book down and think, 'That was interesting.'

Or you can see what I see when I look back over it:

One pattern.

A patterned universe.

Patterned structures.

Patterned bodies.

Patterned thoughts.

Patterned trades.

Patterned lives.

Some of those patterns were given to you. Some you built yourself. Some you inherited without question from family, culture, religion, and school. Some you chose deliberately. Many you didn't.

The invitation in all of this is simple, but not easy:

Stop living as if you are a spectator.

Stop acting as if your beliefs are a side issue.

Stop waiting for someone else—a guru, a priest, a politician, a hedge fund, a secret group—to hand you permission to become who you already sense you could be.

You live in a patterned, responsive universe that is watching, in its own way, how you show up.

You have a brain that will faithfully automate whatever you rehearse—useful or useless, constructive or destructive. You have a body that will echo your habits back to you—either as illness and fatigue, or as energy and resilience. You have a nervous system that will either amplify your best qualities under pressure or collapse into your worst fears, depending on how you've trained it.

You have money, relationships, health and time that will, in the end, be shaped far more by what you do consistently than by what you occasionally wish for.

And somewhere underneath all the scripts and scars, you have a spark—call it Monad, higher self, soul, consciousness, or simply 'the bit of me that knows there must be more'—that refuses to die, no matter how much fear or disappointment you stack on top of it.

From here, the rest is not up to me.

The rest is between you and the life you build.

Maybe you'll take some of the tools here—the wealth mindset, the visualisation, the breathing, the trading structures, the reframes—and quietly weave them into your days. Maybe you'll change how you talk to yourself. Maybe you'll tidy up one corner of your environment so your mind has less clutter. Maybe you'll finally treat sleep like an ally instead of an inconvenience. Maybe you'll stop saying, "I'm just not that type of person," and start saying, "I'm becoming the person who..."

Maybe you do none of it now — put the book down, live ten more years, and only then see what I'm pointing at. Fine. But ask yourself plainly: what do you actually want? Most people can't name it. Identity comes first — you cannot design success if you haven't designed who you are. Success by design starts with a clear answer to that question; without it, commitments fray, and goals remain shadows.

But don't mistake the clarity of that question for soft thinking — searching for who you are can lead you into strange, surprising places. For example, some projects that began as earnest experiments in human potential read like satire now.

If you've ever watched The Men Who Stare at Goats and seen George Clooney trying to kill a goat with his mind, it's easy to file the whole 'First Earth Battalion' idea under comedy or fiction.

Not imagining that governments have, in fact, spent hundreds of millions of dollars on these types of projects.

Just like the CIA material we looked at, the point isn't that every claim was true; it's that these subjects were taken seriously enough to fund, test and document. When the official story looks like a joke, it's usually a sign there's more to it than you're supposed to notice.

Hidden techniques, old programming, and remember your left brain is there filtering it all out and keeping you in autopilot mode. Trusting and believing the politicians.

## Monkey Business

For the fun of it. Go and research the monkey experiment. It has become a kind of legend. Whether every detail is exact or not almost doesn't matter, because the pattern it shows is even evident in your life and is very real.

Imagine five monkeys in a cage.

In the middle of the cage is a set of steps, and at the top of the steps is a bunch of bananas. The monkeys are hungry. Naturally, one of them starts climbing the steps to grab a banana.

The moment his hand reaches out, the researchers hit all five monkeys with a freezing blast of water.

Panic. Screaming. Chaos.

After a while, they calm down. The bananas are still there. Another monkey thinks, "Well, maybe this time..." and starts to climb.

Same thing.

Freezing water. All of them were soaked. All of them were punished.

It doesn't take long before something interesting happens. As soon as any monkey even looks like he's going to climb the steps, the others grab him and drag him down. They don't wait for the water. They beat him up themselves.

They've learned: 'Reaching for bananas = pain for everyone.'

Now the researchers start swapping monkeys.

They take one of the original monkeys out and put a new monkey in. The new guy knows nothing. All he can see is a ladder and bananas.

He goes to climb.

Before he even reaches the top step, the other four monkeys attack him and drag him off the ladder. He has no idea why. He's never been soaked. But he learns fast: "When I go for bananas, I get hurt."

The researchers remove another of the original monkeys and bring in a second newcomer.

Same thing.

He goes to climb, and the group—now including the first new monkey—pile on him. Remember, monkey number one has never seen the hose. He's only ever been beaten up by the others. But he joins in now. He knows the rule, even if he doesn't know the reason.

They keep swapping monkeys until, eventually, there are five monkeys in the cage who have never been hit with water.

Not once.

Yet if you put bananas at the top of the steps, none of them will go near it. And if any one of them dares to try, the others will drag him down. Violently.

If you could interview one of those final monkeys and ask, "Why can't you climb the ladder?" he'd probably shrug and say something like, "That's just how it is. Around here, we don't go for bananas. It's bad."

He wouldn't know about the hose, he wouldn't know why, he'd just know the rule.

That's how most of us are living.

You watched what happened when someone dared to reach: the uncle who started a business and went bankrupt, the neighbour who 'lost everything' in a crash, the person who made money and became the villain in every family story. You felt the emotional 'hose' even if you never saw the original event.

Now, as an adult, when you start to reach for your own 'bananas'—more income, a bigger life, a different career, serious wealth—you can feel something in you pulling back. Or you notice other people trying to drag you down.

Often, they don't know why they're reacting like that. They're the latter monkeys. Nobody's spraying them anymore. They just know, at a deep, unspoken level, 'Reaching = danger.'

The entire purpose of this book was to emphasise that the wealth and freedom you say you want will not arrive as a magic parcel on your doorstep. They will arrive, if they come at all, as the compound effect of thousands of small, aligned choices made by someone who has decided to believe in themselves in a world that quietly benefits when they don't.

The word belief is key here.

It has to be you. No one else can or will do it for you.

You've seen enough now to know that you are not a powerless bystander in a meaningless universe.

You are part of the pattern.

When you read books such as Atomic Habits by James Clear, you know the power that small actions can have, but now, with what you've just read. Now you understand the deeper pattern underneath those habits: identity, belief, state, and signal. When you put the two together, you're no longer just trying to "do better"; you're redesigning who you are, and the life that has no choice but to form around that.

So as you close this book, the question is no longer, "Is there more?"

You already know there is.

The question now is, "What will I do with that knowledge?"

If this was useful, a quick share might help someone else.

# Questions to ask yourself

1. Where in my life do my daily actions contradict what I say I want (wealth, health, freedom, peace)—and what hidden "I am..." belief must be driving that contradiction?

2. When I imagine my ideal life as already real, which part of that vision feels most uncomfortable or "not me"—and what does that reveal about what I currently think I deserve?

3. What sentences about money, success, and "people like us" did I grow up hearing—and how are those sentences still quietly shaping my decisions today?

4. If my subconscious is trying to protect me by keeping my life the same, what does it think it's protecting me from (being seen, failing, succeeding, responsibility, judgment, change)?

5. In which situations do I feel my state drop fastest (money, relationships, risk, criticism)—and what is the first automatic thought or body sensation that shows up in those moments?

6. If my life were a trading account, where do I consistently:

- Cut winners short (quit too early, undercharge, self-sabotage)?

- Let losers run (tolerate bad habits, poor environments, draining people)?

7. Looking at my environment (home, desk, phone, people), what is it training me to expect—scarcity or abundance, chaos or order, distraction or focus—and what one change would shift that signal?

8. What identity label do I cling to most tightly (e.g., "I'm bad with money," "I'm anxious," "I'm not disciplined")—and what benefit do I secretly get from keeping that label?

9. If I treated my thoughts, emotions, and body state as technology I'm programming—not random noise—what is one belief, one habit, and one physical routine I would start updating this week?

10. If I were my own coach or hypnotist, what is the one uncomfortable truth I'd ask myself to face right now—and what new "I am..." statement would I install in its place?

---

yourself. Readers leave with a blueprint: consciously upgrade "I am..." statements, act in alignment with them (especially under pressure), treat setbacks as feedback, and see faith, religion, and self-talk as practical tools for rewriting identity, owning their financial life, and building a resilient wealth mindset.

### Chapter 3 - Visualisation and Success

Explains that wealth mindset is your core identity ("I am someone who builds wealth"), while visualisation is the practical technique for installing that identity by "living in the end" now, not merely wishing for a future goal. It grounds visualisation in science—showing that vivid, emotional mental rehearsal activates many of the same brain areas as real action, can create "real-feeling" memories, and is widely used in sports psychology—so the brain starts to treat your desired future as familiar and achievable. The chapter then gives a clear method: daily, sensory-rich visualisation (often supported by mood boards, affirmations, and "as if" journaling), always followed by at least one aligned action that day. It also highlights the main failure point—people stop when reality pushes back—reframing visualisation as a serious, consistent practice that links belief, brain wiring, and behaviour, and that most successful people used long before their external success appeared.

## Chapter 4 - The Science of Attraction

Grounds the book's earlier "attraction" ideas in practical psychology and science, showing how mindset, language, body, and behaviour mechanically shape results. It explains that we constantly influence reality through micro-signals—facial expressions, posture, tone—and through our internal "code" of words, images, and feelings (NLP), which act as instructions to the brain. Confidence is reframed as visible behaviour that changes how others treat us, while quantum concepts like the observer effect and entanglement are used as metaphors to show how focus, choice, and interconnected systems (mind, body, habits, environment) determine which possibilities become real. "Vibration" is de-mystified as our current physiological and behavioural state, which predictably attracts or repels opportunities. The chapter ties back to wealth mindset, faith, and visualisation from Chapters 1–3, then offers concrete steps—editing "I am..." statements, using body to shift state, daily visualisation, and small consistent actions—framing attraction not as magic but as momentum built from aligned thought, feeling, and action.

## Chapter 5 - Mysteries of Ancient Constructions

Zooms out from personal mindset to show the same core ideas—belief, vibration, alignment, structure—echoed in ancient architecture and lost knowledge. It walks through puzzles like the Great Pyramid's engineering and alignment, "impossible" global stonework, and evidence that past cultures deliberately used geometry, sound, and resonance as a kind of technology, not just decoration or tomb-building. This wider view suggests patterns repeat across shells, charts, temples, and cathedrals; powerful knowledge can be hidden or forgotten; and function is often wrapped in story, just as practical mind-tech hides inside religion and myth. The chapter's punchline is that your thoughts, emotions, and "I am" stories are not fluffy extras but precise tools—inner technologies you can use, with as much intention as ancient builders used stone and sound—to shape your life, wealth, and trading results.

## Chapter 6 - Frequencies

Turns "frequency" into something concrete and practical. It explains that you are literally made of vibrating structures, and your emotional state (fear vs calm, scarcity vs confidence) is a frequency pattern that changes what you notice, how you act, and what you experience. Sound and brainwave research—plus ancient practices like chanting and "sleep temples"—show that tones and rhythms can reliably shift state, just as different brainwaves support sleep, learning, focus, or peak performance. Crystals then anchor the idea that frequency + structure = function: quartz's ordered lattice and piezoelectric effect let it convert, stabilise, and store energy and information, which is why both ancient cultures and modern tech use it. This becomes a metaphor for mindset and trading: your beliefs and habits are patterns that store "data" and determine how you respond under pressure. By repeating empowering "I am" statements, visualising your future with emotion, and surrounding yourself with aligned cues, you "crystallise" a coherent inner signal. The chapter's core lesson is that success isn't random—it's built by deliberately shaping the structures (beliefs, habits, environments) that hold the frequency you want to live in, until you become a stable, clear signal amid noise.

## Chapter 7 - Technology, Religion, & Numbers

Shows that technology, mathematics, and spirituality are likely different faces of the same underlying pattern—and that this pattern directly applies to trading and personal agency. Every technology begins as an idea that's turned into numbers and logic (circuits, code), just as your beliefs and habits form the "code" that produces your results. Simple rules plus repetition (Mandelbrot, Fibonacci) create infinite complexity in nature and markets, suggesting reality is patterned rather than random. Mathematics becomes a bridge between "God" and "code," with both scripture and physics pointing to ordered information or logos. As AI and quantum computing advance, science and spirituality increasingly overlap, while trading clearly shows that good maths (a solid system) is necessary but useless if the psychological "operating system" is broken. The chapter ends by asking you to see your thoughts as code, your ideas as blueprints, and markets as structured systems—inviting you to work with the pattern, not against it, and to recognise that you are already programming your own life and trading, whether consciously or not.

## Chapter 8 - Motivation and Stoicism

Bolts a spine onto everything so far by showing that mindset without resilience is just a mood, and that motivation only becomes reliable when it's backed by Stoic discipline. It explains that most limits are installed, not real, and that real motivation is knowing your why and moving especially when you don't feel like it. Stoicism is presented as emotional risk management: focus only on what you can control, rehearse setbacks in advance, and treat daily annoyances (burnt toast, bad emails, losing trades) as training drills rather than proof your day is ruined. This steadiness under pressure is what keeps your "frequency" clear and consistent, so wealth mindset, faith, and visualisation don't collapse at the first sign of friction. The core message: your real edge is who you choose to be on bad days; when you combine Stoic stability with a genuine wealth mindset and trained imagination, you stop being a passenger and become someone the universe can take seriously.

## Chapter 9 - Mind and Body Connection

Shows that success is a physiological project as much as a mental one: mind and body are a single system, so thoughts, emotions, hormones, nerves, and immunity are always influencing each other. Neuroplasticity means every repeated thought or feeling trains your brain, while emotions live in the body and can be changed via posture and breath. Placebo/nocebo prove belief alters flesh; your body is the hardware and your mind the operating system, so both must be maintained. Food, movement, sleep, meditation, breathwork, and embodied practices become non-negotiable mindset tools, clearing "static" so affirmations and visualisation can actually take root. Emotional intelligence—reading and regulating your own and others' states—emerges as a core wealth skill. The chapter's message: when you align physical foundations with mental training, you create a self-reinforcing loop where better biology supports better decisions and confidence, and you stop living like a pawn and start playing the game of life deliberately.

### Chapter 10 - The Hidden Circuit

Intricately connects tai chi with personal development and trading, emphasising the mind-body connection and the transformative power of intention and energy flow. Initially dismissing tai chi as mere "retirement-home yoga," the author's evolving perspective reveals its profound capacity for grounding and focus, paralleling essential principles in trading. Key concepts like "rooting," "centre," and "song" are introduced to highlight how internal stability affects external outcomes. The chapter explores energy manipulation through mudras and Dim Mak, contrasting gentle practices with more aggressive techniques while reinforcing the notion that our physical actions influence our mental states. Ultimately, the chapter invites readers to consciously train their internal state to respond effectively under pressure, blending ancient wisdom with modern applications for personal growth.

## Chapter 11 - Hypnotism & Drugs

Shows that altered states of consciousness—via hypnosis, psychedelics, deep meditation, astral projection, and ancient spiritual practices—are all different ways of accessing and influencing the subconscious, which quietly runs most of your life. It separates hype about "mind control" from hypnosis as focused, relaxed attention that lets you speak to deeper beliefs, fears, and identity patterns in a language the subconscious understands: calm, vivid, emotionally rich imagery and "living in the end." The chapter positions astral projection as a more spiritual exploration and hypnosis as more therapeutic, but both sit in the same family of human attempts to go beyond ordinary awareness. Practically, it invites you to treat hypnosis as self-programming, to use simple scripts and targeted inner questions to uncover and change limiting patterns, and to recognise that when prayers, affirmations, or the Law of Attraction seem to fail, the real issue is usually misaligned inner code—not lack of will.

## Chapter 12 - Self-Imposed Limits

Explains that your biggest barriers are not external but the invisible ceilings you've accepted about who you are and what you're allowed to have. It shows how repeated messages ("be realistic," "people like us don't get rich") become hypnotic "I am..." scripts that the subconscious treats as instructions, quietly shaping behaviour, sabotaging goals, and capping wealth. Hypnosis and similar tools are reframed as ways to uninstall and upgrade these scripts, revealing the Monad—your deeper, unconditioned self—beneath fear and doubt. Practically, the chapter teaches you to expose and rewrite limiting narratives, flip "I am" statements into growth-oriented identities, use trance and repetition to embed them, act "as if" in small daily ways, and regularly connect with your deeper self. The core message: real freedom comes from expanding the identity you live from; as that inner container grows, your capacity for wealth and fulfilment grows with it.

### Chapter 13 - Community Influence and Personal Growth

Explains that most gurus, coaches, and "mindset experts" are repackaging the same core principles—belief, identity, responsibility, and action—using different styles and stories, and that the real issue is usually not missing information but missing integration. It explains how events and communities can genuinely help by shifting your environment and normalising success, but can also create dependency and guru-worship if you outsource your power and identity to them. The chapter teaches you to filter all content through basic principles, watch whether you feel more empowered or more needy afterward, use communities strategically (for accountability and feedback, not identity), and treat mentors as partners rather than saviours. Above all, it reframes you as your own best coach: you already have the tools—reframing beliefs, visualisation, state management, habit work—and the real "secret" is consistently applying them and becoming the primary influencer in your own life.

## Chapter 14 - Deeper than you realise

Connecting everything in the book so far—money, trading, religion, geometry, hypnosis, mindset—is pointing at the same core truth: reality is far deeper and more patterned than we were taught, and you are an active participant in it, not a passive spectator. By linking pyramids, quantum experiments, Akashic-style ideas, symbols, self-help frameworks, and psychology, the chapter argues that no guru or system owns "the secret"; they're all circling the same principles: thoughts and beliefs shape perception and action, mind and body are one system, symbols and structures carry meaning, and your identity scripts quietly create your ceiling. The reader learns to stop chasing missing pieces and instead use what they already know—visualisation, "I am" statements, state management, mind–body care, environment design, and consistent action—to interact consciously with this deeper pattern. The chapter leaves you with a clear stance: you don't need to understand everything to live differently; you only need to accept that life runs deeper than the surface story and start playing as the mover of the pieces, not just a piece on the board.

## Chapter 15 - CIA

Shows that many ideas often dismissed as "woo-woo" altered states, brainwave entrainment, out-of-body experiences, holographic reality, consciousness as a field—have been taken seriously enough by governments and intelligence agencies to study in depth. Declassified documents on the Gateway Process, MKULTRA, remote viewing, and Tesla's seized papers reveal that major institutions have explored how sound, suggestion, drugs, and state shifts might steer perception, extract information, or alter behaviour. This doesn't prove every wild claim, but it strongly suggests your mind is far more malleable and powerful than you were taught. The chapter invites you to treat beliefs, "I AM" statements, visualisation, and emotional state as operational tools—not fluff—reinterpreting wealth mindset and state management as interfaces with a deeper field, practising "energetic hygiene" with your inputs, and dropping the idea that thoughts are powerless.

### Chapter 16 - Sacred Geometry

Showing that reality is structured by recurring patterns, ratios, and shapes that appear in everything from pyramids, cathedrals, and Masonic symbols to shells, galaxies, cymatic sound patterns, and trading charts. It argues that these aren't mystical decorations but evidence of a universe that prefers balance, proportion, and rhythm—and that our minds, bodies, spaces, and finances work best when they mirror those qualities. Practically, the chapter invites you to use geometry as a lens: design ordered environments to support clear thinking, balance key "ratios" in your life (work/rest, risk/safety, learning/integration), see charts as human geometry rather than chaos, and treat your growth like architecture—foundations, pillars, then arches—so your life and wealth align with the deeper patterns reality already runs on.

## Chapter 17 - Different Meditation

Explaining that meditation isn't about forcing yourself to "do nothing," but about training the simplest, hardest skill: staying with one thing long enough for it to change you. Using the Zen ideas of Muso no Ichi (becoming one with the action) and Ichigyo-Zammai (one-action samadhi), the chapter reframes focus as a muscle you can deliberately build through total immersion in a single task—whether that's a breath, a chart, an email, or brushing your teeth. It links this to flow state, deep work, and earlier themes like Neville's "living in the end" and the observer effect: when you reduce distraction and return, gently but relentlessly, to one chosen object, self-talk quiets, fear loses its grip, and you gain a stable inner place to stand. The reader learns how to turn ordinary moments into practice, create the conditions for clean, effortless action in high-stakes situations, and use meditation not as an escape from life but as a practical way to design it—moving from being yanked around by thoughts and notifications to acting as the player, not the pawn.

## Chapter 18 - Other forms of Psychology

Chapter 18 shows that the real edge in trading, business, and wealth has little to do with strategy and everything to do with the "code" of beliefs, identity, and emotional conditioning running in the background. Using the contrast between students on the same system with opposite results, it introduces self-actualisation (Maslow), CBT, schema therapy, and ACT to explain how deep "I am..." stories and early schemas shape how traders interpret losses, whether they adapt or sabotage, and whether success even feels safe. The chapter then zooms out to music and media—Bach, Mozart, Beethoven, Wagner, film scores, Jaws, Star Wars, Titanic, Avatar, and algorithmic feeds—as proof that humans are programmable by rhythm, story, and repetition at a bodily level. It argues that these cultural "soundtracks" often link wealth with villainy and struggle with virtue, quietly sabotaging your wealth mindset, and calls you to become your own psychologist and programmer: seeing your patterns clearly, separating money from morality, installing new, ethical models of wealth, and using tools like "I am" statements, visualisation, breath, and environment to rewrite your inner score so your external life can finally change.

## Chapter 19 - Connecting the Dots

A concise, unifying finale that pulls the book's threads into a single, practical truth: patterns underlie money, mind, ritual, science and habit, and belief—your self-script—drives outcome. Moving from personal memoir to classroom results to neuroscience, ritual and odd historical curiosities (pyramids, cymatics, the monkey experiment) to show how repeated attention shapes state, behaviour and results. Clear, urgent, and action-focused, the chapter reframes mindset work as identity design: change the inner script, use community and tools as scaffolding, and you expand what's possible in life and wealth.

Overall the chapter functions as an effective capstone: clarifying the book's through-line, inspiring responsibility, and leaving the reader both challenged and equipped to act.

## Success by Design – Work With Me

If you've read this far, you already know this book was never just about trading or pyramids.

It's about how you think, how you move, how you make decisions under pressure—and how all of that quietly builds (or destroys) your results.

Over the last 25+ years I've been:

- A trader and investor
- A founder and inventor
- A non-executive director
- A consultant and coach to business owners, traders and professionals

Underneath all those titles, it's the same job:

Helping people and companies find the hidden pattern that's running them—and then redesign it on purpose.

It could be your personal life not just a business.

If you want help with that, there are a few ways we can work together.

## 1. Private Consulting & Coaching

I work selectively with:

- Start-ups
- Founders and business owners
- Traders and investors
- Professionals in transition
- Boards looking for a non-exec with a real-world operator's view

Typical work includes:

- Clarifying strategy and simplifying complexity
- Diagnosing the "hidden code" behind recurring problems
- Designing systems (business and personal) that actually fit how you think
- Aligning mindset, behaviour and numbers so growth is sustainable

If you'd like to explore working together, email:

**welshexpertgeneralist@gmail.com**

Include a brief outline of who you are, what you're working on, and what you think you need. If I'm not the right fit, I'll tell you straight.

## 2. Join the Community

If you want to be around people who think like this—who care about wealth, mindset, freedom and doing things properly—you're welcome in our Discord.

Inside, we have a dedicated **"success-by-design"** channel where readers of this book and members of our trading community:

- Discuss the ideas you've just read
- Share practical applications and experiments
- Hold each other accountable as they implement

You don't need to impress anyone. You just need to be serious about changing your own code.

You can find the link to join at:

**Website - https://www.lewis-daniels.com/**

**Discord - https://discord.gg/5t3w47raHe**

---

### 3. Speaking & Advisory

I also advise and speak for:

- Companies wanting to integrate mindset and performance into their culture
- Investment and trading groups needing a blunt, real-world perspective
- Events focused on high performance, wealth, and unconventional thinking

For enquiries, use the same email:

**welshexpertgeneralist@gmail.com**

---

Whatever route you choose—private work, the community, or simply taking these ideas and running with them on your own—remember this:

You are not short of information.

What you've been short of, until now, is a way to join the dots and design a life that actually matches who you know you could be.

www.ingramcontent.com/pod-product-compliance
Lightning Source LLC
Chambersburg PA
CBHW050521240926
51846CB00003B/645
* 9 7 8 3 9 1 6 9 7 1 5 1 0 *